23rd Topical Conference on Refinery Processing 2020

Topical Conference at the 2020 AIChE Spring Meeting and 16th Global Congress on Process Safety

Online
17 – 21 August 2020

ISBN: 978-1-7138-2326-1

Printed from e-media with permission by:

Curran Associates, Inc.
57 Morehouse Lane
Red Hook, NY 12571

Some format issues inherent in the e-media version may also appear in this print version.

Printed with permission by Curran Associates, Inc. (2021)

For permission requests, please contact AIChE
at the address below.

AIChE
120 Wall Street, FL 23
New York, NY 10005-4020

Phone: (800) 242-4363
Fax: (203) 775-5177

www.aiche.org

Additional copies of this publication are available from:

Curran Associates, Inc.
57 Morehouse Lane
Red Hook, NY 12571 USA
Phone: 845-758-0400
Fax: 845-758-2633
Email: curran@proceedings.com
Web: www.proceedings.com

23rd Topical Conference on Refinery Processing 2020

Topical Conference at the 2020 AIChE Spring Meeting and 16th Global Congress on Process Safety

Online
17 – 21 August 2020

AIChE
The Global Home of Chemical Engineers

TABLE OF CONTENTS

591161 2020 AIChE Spring Meeting & 16th GCPS AGILE Award Keynote Address - Beyond the Horizon: The Factors and Forces Affecting the Global Petrochemical Industry

Abstract:

Over the past 10 years, LyondellBasell and the global petrochemical industry has undergone a dramatic transformation. In this era of unprecedented disruption, it is likely that the next the decade will present both challenges and opportunities for the petrochemical sector. LyondellBasell CEO Bob Patel will summarize the company's progress to date and provide a forward-looking perspective on key opportunities and risks faced by the industry in 2020 and beyond.

592550 Digital Transformation in Chemical Industry

Abstract:

Manufacturing is the original denizen of one of the cornerstones of Industry 4.0 - a digital transformation journey to unlock the full potential of manufacturing data. This talk will highlight the critical human element to deliver sustainable and scalable digital transformation, discuss common challenges using Artificial Intelligence and Advanced Analytics as an example of digital transformation, focusing on corporate culture change. A proven best practice will be shared to illustrate real-world outcome in digital transformation.

591157 Securing Enterprises via Cybersecurity Understanding, Resiliency, and Education

Abstract:

Securing Enterprises via Cybersecurity Understanding, Resiliency, and Education

Industry 4.0 initiatives are enabled by leveraging the deployment of information technologies into the operational technology environment. While a logical approach in defining the future of smart manufacturing, it remains fret with challenges and risks that must be mitigated assure enterprise resilience to cyber and cyber-physical attacks. As economies recalibrate in the aftermath of the current public health emergency, definitions of what constitutes "critical infrastructure" are broadening while, at the same time, attack surfaces and the threat landscape are expanding. This means companies must place a greater emphasis on taking the steps and making the investments necessary to overcome the challenges and navigate these challenges and risks to maximize their resilience and secure and protect their enterprises. This presentation will explore the link between cybersecurity and process safety in the context of a safety culture. It will also identify cybersecurity challenges in industry 4.0 initiatives, the integration of information technologies (IT) and operational technologies (IT), the delivery of training, and the need for credentialing and ongoing professional development.

582984 Key Note: Heat Exchanger Design on Paper Vs Practice

Abstract:

Shell-and-Tube heat exchangers in the Oil & Gas business are designed using Standards which cover mechanical, thermal, and fabrication aspects. Multiple Standards are used; Industry Standards such as API, ISO, TEMA and internal Standards of the EPC and end-user companies. The Standards are based on many years of combined experience of the several companies involved in the design, fabrication, and operation of the heat exchangers. The purpose of the Standards is to ensure a minimum quality in the design, especially when the design may be done by a party without access to operational experience. In the field, heat exchangers often operate at flow, temperature, pressure, and fluid composition conditions substantially different than assumed in the design. The purpose of this presentation is to briefly discuss the philosophy behind Standards and to highlight situations where Standards do not take into account operational variations. These variations from design assumptions lead to unexpected failures of heat exchangers. The presentation will emphasize the point of view that although all Standards are based on sound principles, the designer must consider operational issues and deviate from the Standards as much as necessary to create a reliable design for real life.

HELIXCHANGER® Heat Exchanger

*R. J. Jibb[1], S. Taha[1], G. Podrebarac[2]
[1] Lummus Technology LLC*, Bloomfield NJ
[2] Lummus Technology LLC, Pasadena,

Abstract

Case studies spanning 20 years of experience with Lummus Technology's HELIXCHANGER[1] heat exchanger provide a body of evidence that this technology can be used to improve performance, increase run length and reduce operating expense (OPEX) in a number of applications. A new case study is presented comparing HELIXCHANGER and conventional segmental bundle performance for cooling of a viscous Sales Oil. Laboratory testing of HELIXCHANGER bundles provides insights into the mechanisms involved, and a basis for future predictive models for this type of equipment. Additionally, laboratory testing of HELIXCHANGER sealing arrangements recommended for segmental baffles reveals that they may not be optimal for HELIXCHANGER helical baffles, and research is ongoing to develop sealing arrangements more appropriate for a helical flow pattern. Some promising initial results are presented.

Introduction

The term "Helical Baffles" refers to a class of baffle structures used in shell and tube heat exchangers to direct fluid on the exterior of a tube bundle in a primarily helical path around a central core. This contrasts with flow in shell and tube heat exchangers with cross-baffles (predominantly crossflow, perpendicular to the tubes) and "grid" or "rod" baffles (predominantly longitudinal flow, parallel to the tube axis).

HELIXCHANGER® Heat Exchanger

With over 4000 units in operation worldwide, the most widely used helical baffle geometry in industrial applications is the HELIXCHANGER[1] heat exchanger which is shown in Figure 1, and described in detail in a number of references [1-5]. The HELIXCHANGER technology was developed from the early design of Lutcha and Nemcansky [6] in The Czech Republic and later acquired by Lummus Heat Transfer. The design was improved further as described in US Patent 6827138 [7].

Figure 1 HELIXCHANGER bundles during fabrication

---

[1] *HELIXCHANGER® heat exchanger is a proprietary shell and tube heat exchanger technology licensed by Lummus Technology LLC*

Referring to **Figure 2**, some key features of the HELIXCHANGER baffle design are:

a) Quadrant shaped baffles positioned at an angle to the tube axis along a central rod.
b) Baffles act as guide vanes rather than flow barriers.
c) Baffles are cut from an *elliptical plate* so that the clearance between the baffle and the shell is consistent.
d) Baffle quadrants *overlap* which enables a balance between the rotational component in the outer region and the higher velocity vortical flow in the center of the bundle. Correct selection of the *baffle angle*, and *overlap* will yield a uniform temperature profile and minimize axial dispersion.

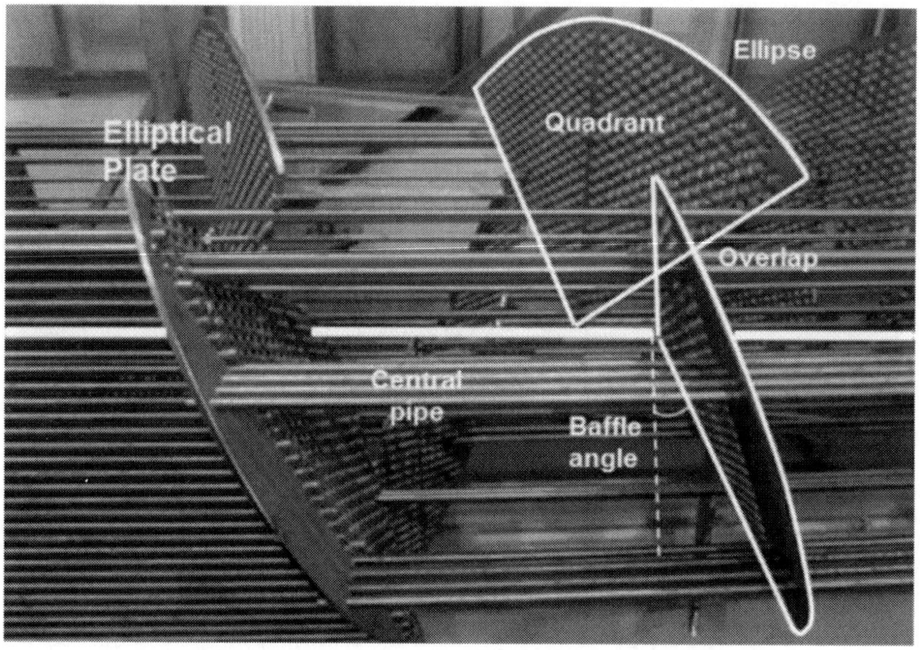

Figure 2 HELIXCHANGER baffle layout (tubes not installed)

For a given shell diameter, helical flow can provide a higher average velocity than crossflow type segmental baffles whilst maintaining the similar pressure drop. This can be an important consideration in many fouling applications where high shear stress on the external tube surfaces is desirable to limit fouling [8], or for viscous fluids where higher shear rates can significantly improve performance. The helical baffle construction also provides a shorter unsupported span than segmental baffles which in turn reduces the risk of flow induced vibration at higher velocities. Where vibration concerns or available shell-side pressure drop are limiting baffle angle, spacing and external sealing can be varied in order to maximize shell-side velocities.

The applications of the HELIXCHANGER technology include the following:

1. ***Bundle retrofits*** – Helical bundles are often used in retrofits where available pressure drop is limited and bundle diameter is constrained by an existing shell. HELIXCHANGER bundles are manufactured to TEMA standards to directly replace TEMA type E and J type bundles.

2. ***Fouling services*** – Where fouling fluids are placed on the shell side, the helical flow pattern has been shown to reduce fouling rates [1, 2, 8, 9] through combination of *lower tube surface temperature* and an *increase in the shear stress over the tube surface*. The continuous helical flow pattern reduces dead zones and back mixing that may be unavoidable with segmental cross-baffles.
3. ***Flow induced vibration*** – The unsupported span of a helical bundle is less that a corresponding segmental baffle design. A higher shell side velocity can often be accommodated without exceeding vibration limits. A double helix design may be applied to provide additional tube support whilst maintaining a relatively large baffle angle.
4. ***HELITOWER™*** – Helical bundles for large vertical feed-effluent heat exchangers such as in reformers and Olefin Conversion Technology (OCT) under the umbrella term *HELITOWER* provide additional heat transfer surface area within the same shell diameter, as well as fouling and vibration benefits, compared to "No tubes in window" (NTIW)
5. ***Viscous fluids*** – HELIXCHANGER overlapping baffle design rotates and divides the flow, in a similar way to a static mixer device to more effectively heat or cool viscous fluids, or fluids which have a large range of viscosity with temperature. This application will be covered in detail in later sections.
6. ***Tube-side enhancement*** – Tube inserts, twisted tapes and other devices may increase tube-side heat transfer coefficient significantly. HELIXCHANGER may then be used on the shell-side to accommodate a resultant higher shell-side velocity (due to smaller tube bundle diameter or reduced shell count) without excessive pressure drop [10].
7. ***Cooling water on shell-side*** – HELIXCHANGER construction can typically enable a higher shell-side velocity which is desirable when it is necessary to place cooling water on shell-side.
8. ***Condensers*** – HELIXCHANGER baffles can be used for shell-side condensing applications. If flow induced vibration concerns are limiting, Helical Baffles can provide a more economical design. The HELIFIN® trade name is used for helical baffles combined with low-fin tubes. HELIFIN units are applied in condensers to maintain a shorter unsupported span increasing reliability against flow induced vibration while lowering the overall capital cost.

There are a number of published reports which provide case studies and operational feedback for the applications listed above.

Master et al. [1, 2] have presented several case studies covering refinery applications in crude preheat and feed-effluent services. In one example, an increase in run length from 1 year between cleanings to over 2 years in *crude preheat* service is reported. Since that time the use of fouling monitoring systems has become more widespread and more data has become available.

Jibb et al. [9] have presented three case studies which illustrate improvements in run length for heating heavy fouling fluids such as *vacuum residue*. Significant increases in run length are again reported.

Gough et al. [10] have provided a case study for grass roots design case study combining HiTRAN® and HELIXCHANGER technologies for a crude oil interchanger on a floating production storage and offloading ship platform refining process. The advantages were smaller plot space at lower energy cost with reduced fouling tendency as compared to conventional exchanger design. The helical flow velocity was more than three times higher than the average cross flow velocity achieved in the equivalent segmental-baffle design.

In this paper we focus on viscous cooling applications. Shell-side cooling of a viscous fluid is desirable due to potentially higher heat transfer performance than tube-side cooling. However, the shell-side geometry is inherently more complex, so the potential for unexpected operational issues arising from maldistribution is higher and the performance can be more difficult to predict. Rosenhow et al. [11] have classified maldistribution according to various root causes. For viscous cooling in shell and tube heat exchangers, the two relevant root causes are – (i) *Geometry induced maldistribution* and (ii) *Flow maldistribution induced by operating conditions* (in this case *viscosity-induced*).

> *Geometry induced maldistribution* occurs as a result of the heat exchanger flow geometry, fabrication, and clearances etc. For a shell and tube heat exchanger this includes factors such as baffle design, sealing strips and clearances provided to allow bundle insertion and removal. The maldistribution is not significantly modified or improved by changing operating conditions.

> *Viscosity-induced maldistribution* is a process whereby operating conditions inevitably lead to combinations of physical properties that cause operating instabilities. For example cooled viscous flow can lead to plug flow in the colder sections of the bundle, with high axial leakage flows around the warmer bundle to shell gap.

A case study is presented which provides a direct comparison of segmental baffle and HELIXCHANGER performance for cooling of a viscous fluid. The later sections describe an experimental study of viscous cooling carried out at the HELIXCHANGER test facility in Pasadena Texas.

Case Study: Sales Oil/BFW Exchanger

A North American operator processing viscous sales oil ran a series of trials to compare the performance of existing conventional segmental baffle heat exchangers against retrofitted HELIXCHANGER heat exchangers installed in parallel. Lummus Heat Transfer (LHT) received a request to provide HELIXCHANGER bundles to replace existing segmental baffle exchangers which were underperforming at the customer's facility. The new HELIXCHANGER bundles were built by a licensed fabricator and installed in 2018. A comparison between the existing bundle geometry and the HELIXCHANGER retrofit bundles is show in Table 1.

The exchanger service was a viscous cooling application with sales oil on the shell side and boiler feed water (BFW) on the tube side. The sales oil bulk viscosity ranged from about 20 cP at the inlet to about 300 cP at the outlet. The unit consisted of four shells and four bundles arranged as two top bundles and two bottom bundles (all identical). 1 top and 1 bottom shell were retrofitted with two identical HELIXCHANGER bundles with the same overall dimensions and nearly identical surface area (Table 1), thus enabling a meaningful comparison between HELIXCHANGER and segmental baffle performance. HELIXCHANGER and Segmental units had the same overall dimensions, but the HELIXCHANGER bundles had slightly less tubes due to construction requirements.

Table 1 HELIXCHANGER and Segmental bundle geometry.

		Segmental	HELIXCHANGER
Number of shells	-	2 shells in series	2 shells in series
Size	in	23 x 22	23 x 22
Tube OD	in	3/4	3/4
Tube pitch	in	1	1
Tube layout angle	deg	45	90
Tube Count	-	162 U's	159 U's
Area	ft2	1156.5	1133.2
Baffle type	-	V-Single Segmental	Single Helix

Previous operating experience (all conventional segmental baffles) indicated that addition of light hydrocarbon product (condensate) or "diluent" was often necessary to achieve an acceptable cooling rate, due to the high viscosity of the sales oil at low temperatures. Additionally, the heat transfer resistance was controlled by the shell-side fluid flow resulting in relatively low tube wall temperature and a high ratio of tube wall viscosity to bulk viscosity. For cooling of viscous fluids, this condition can result in larger bypass flow around the tube bundle and subsequent thermal underperformance [11]. The flow split between the two sections and diluent addition was varied between the two parallel trains and the cooling performance was recorded. The case study therefore has the interesting property that both geometry induced factors (baffle design) and operational factors (diluent) are varied in the course of the test. Figure 3 below summarizes the data obtained from these trials.

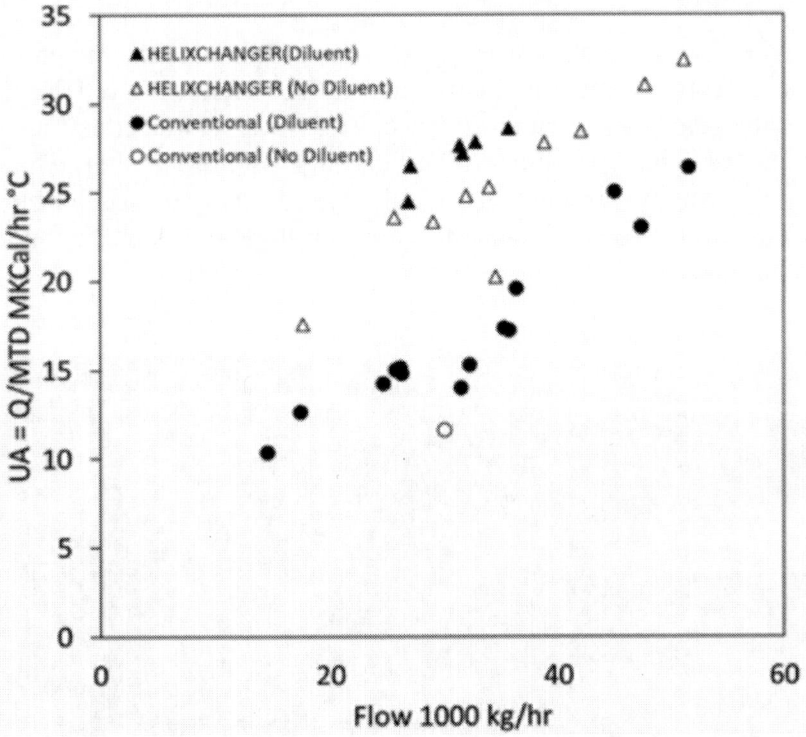

Figure 3 Apparent UA for HELIXCHANGER compared to conventional Segmental baffles as a function of Sales Oil flow and diluent addition

The data in Figure 3 is presented in terms of apparent UA where:

$$UA = \frac{Q}{MTD} \quad (1)$$

The apparent UA of the conventional segmental heat exchanger was in the range 10 – 26 x 10^3 kcal/hr. °C, whereas for the HELIXCHANGER section the range was 17 – 33 x 10^3 kcal/hr. °C. Diluent was added to the conventional section for all runs except one. For the HELIXCHANGER section the addition of diluent resulted in a 5-10% increase in apparent UA over the range considered.

Based on these observations the HELIXCHANGER bundle apparent UA (with diluent) improved by as much as 30 - 50% compared to the parallel section and up to 100% without diluent – although the latter comparison is based on a single data point for the conventional section. A "switching" effect or instability was encountered in the conventional bundles whereby a sudden drop in heat transfer performance occurred in the lower bundle unless a diluent light hydrocarbon product (condensate) was added to the inlet heavy oil stream. The switching behavior did not occur with the HELIXCHANGER bundles. Overall the heat duty was increased by 15-22% which enabled an additional 10-12 °C of sales oil cooling.

HELIXCHANGER test facility

A HELIXCHANGER experimental test facility was commissioned in 2013 to gather performance data for a range of flow conditions. The facility has previously been used to gather test data for pressure drop at high Reynolds numbers in support of Computational Fluid Flow Study to improve pressure drop prediction for this service [12].

More recently, a program was carried out to gather data for cooling of viscous fluids and to investigate the effects of sealing the bundle to shell gap. Two HELIXCHANGER bundles were constructed for the study as shown in Figure 4. The first (A) was constructed as a TEMA type BES with two tube-side passes. This bundle had relatively large bundle to shell gap and a smaller baffle angle of 7°. The second (B) was constructed as a TEMA type BEU with smaller bundle to shell gap and a baffle angle of 12°. The bundle to shell clearance for Bundle B was small enough that it would not normally be considered necessary to employ seal strips.

Figure 4 – Left - Bundle A (TEMA type BES) and (b) Bundle B (TEMA type BEU); Right – Instrumented test exchanger.

Both test bundles were used to investigate the effects of various sealing arrangements on the heat transfer, pressure drop and flow distribution over a wide range of flow conditions. The test fluids included water, and various fluids of low, medium and high viscosity including Newtonian and non-Newtonian (shear thinning) fluids. The cooling fluid on the tube-side was chilled water.

The two bundles were tested with one of three sealing configurations:

1. **"No Seal"** – No sealing device at all
2. **"4 strips"** – Axial seal strips at 4 locations
3. **"New Seal"** – Patent pending seal strip design for helical flow.

Measurements of overall heat transfer rate and pressure drop were supplemented with intermediate temperature and pressure measurements and a residence time measurement using an optical technique. The latter was used to determine the experimental value of Peclet Number (*Pe).*

The Peclet number and dispersion coefficient were calculated after Roetzel and Balzereit [13] using a method presented by the American Institute of Chemical Engineers, Dudukovic [14]. First, the mean residence time and variance for the inlet and outlet response curves were determined. The mean residence time and variance of the HELIXCHANGER were calculated by taking the difference of the inlet and outlet residence time and variance (Equations 2 and 3). The Peclet number was calculated as shown in Equation 4.

$$\mu = \mu_{out} - \mu_{in} \quad (2)$$

$$\sigma^2 = \sigma^2_{out} - \sigma^2_{in} \quad (3)$$

$$Pe = \begin{cases} \frac{2\mu^2}{\sigma^2} & \frac{2\mu^2}{\sigma^2} > 10 \\ \frac{\mu^2+\sqrt{\mu^2(\mu^2+8\sigma^2)}}{\sigma^2} & \frac{2\mu^2}{\sigma^2} < 10 \end{cases} \quad (4)$$

Where, in this case, μ is the mean residence time (not viscosity), and σ^2 is the variance.

It should also be noted that the Reynolds number reported by Roetzel and Balzeriet [13] is not the same as the helical Reynolds number ($Re_{helical}$) used in later analysis. The former is calculated using a method from *VDI-W rmeatlas*, whilst the latter is calculated based on tube outside diameter, with a proprietary method to determine the fluid velocity for the helical fluid path.

Roetzel and Xuan [15] and later Roetzel and Lee [16] have shown experimentally that *Pe* for a shell and tube heat exchanger depends strongly on the shell to baffle clearance (STB) and can vary from 5 (at 2.5 mm STB clearance) to over 50 (for zero STB clearance). For reference the current test units were fabricated using (nominal) TEMA STB clearances of 1.6 mm, although later measurements revealed that the as-built value was somewhat larger (around 2.3 mm).

An alternative method for determining the effectiveness of a shell-side flow configuration is the so called "pressure drop conversion factor" (C_{pd}) defined by Palen and Taborek [17], which is derived based on the Colburn analogy. This method is documented in HTRI reports and so is not reproduced here.

The parameter C_{pd} provides a measure of the efficiency with which pressure drop is converted to heat transfer. It has the important property that it accounts for the bypass and leakage effects inherent to shell-side flow without prior knowledge or estimation of the B-stream fraction as defined by Tinker [18]. If the leakage/bypass is relatively large there will be a lower pressure drop but correspondingly also a lower apparent heat transfer coefficient. Moreover, C_{pd} can be determined directly from pressure drop measurements given a knowledge of the exchanger geometry. The values for C_{pd} may be determined from experimental or field measurements. Bouharie [19] has used the pressure drop conversion factor C_{pd} to compare various seal strip configurations for segmental baffles, and again this information is available from HTRI.

For a HELIXCHANGER, C_{pd} is defined based on a single helical lead (one single rotation), so the relevant parameters are calculated according to this definition. According to Palen and Taborek [17], the theoretical maximum value of C_{pd} for pure crossflow with no leakage or bypass effects is ~62.5% (for turbulent flow) and ~25% (for laminar flow). By comparing the measured value of C_{pd} to the theoretical maximum it is possible to determine the effectiveness of a given bundle configuration. Additionally, it can be used to determine the relative performance of different sealing arrangements in converting pressure drop to heat transfer.

The two derived parameters (C_{pd} and Pe) are independent measurements. By measuring both factors concurrently, it is possible to understand the improvement in heat transfer and pressure drop performance for shell-side flow in a more complete way than can be achieved by either method alone.

High viscosity fluid tests ($30 < Re_{helical} < 1,000$)

Figure 5 shows the parameters C_{pd} and Pe derived from the test data as a function of the helical Reynolds number $Re_{helical}$. For shell-side Reynolds numbers around 1000 Bundle A showed a 13% increase in heat transfer for the sealed case (4 strips) with a pressure drop penalty of 20%. Measurements of Pe were in the range 10 to 12 and were not affected by the addition of seal strips. At Reynolds number from 1000 down to 60 both C_{pd} *and* Pe decreased from about 38% to 15% and from 12 to 8 respectively, as the effects of flow bypass and leakage became more prominent. Again, the addition of seal strips had little or no discernible effect on either parameter.

For very low shell-side Reynolds numbers in the range 50 - 150, large temperature gradients were observed in all cases. For Bundle A the conventional seal increased the heat transfer coefficient by ~14% at a pressure drop penalty of 36%. However, C_{pd} for the sealed and unsealed cases were comparable in the range 15 – 24%.

A reduction in both C_{pd} and Pe at lower Reynolds numbers was also observed for Bundle B. For Reynolds numbers from 100 to 1000 a slightly lower value of C_{pd} was observed for the runs with seal strips than those without seal strips. Measurements in this range support the existing design practice which would not require seal strips for the BEU bundle.

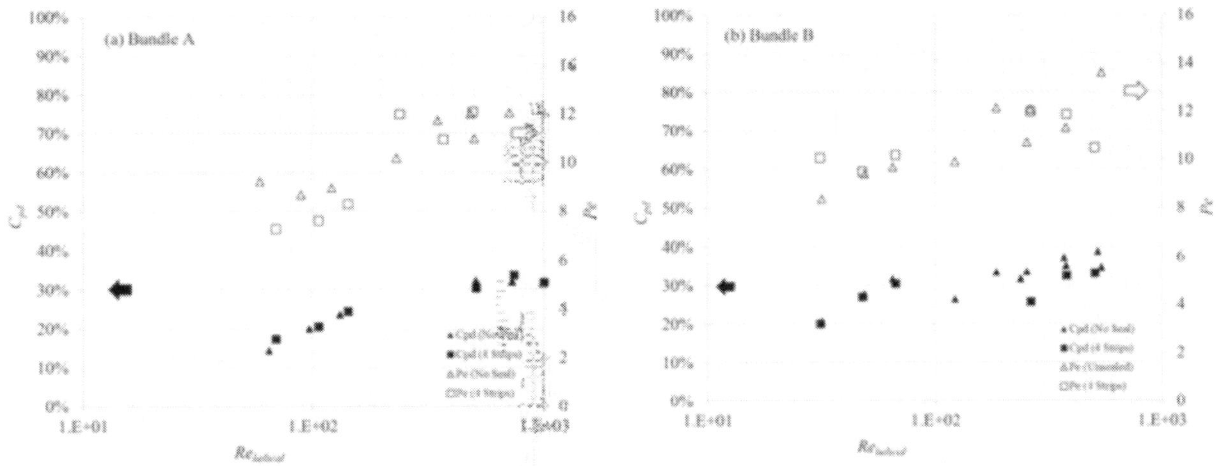

Figure 5 *Pe* and C_{pd} values derived from laboratory measurements (a) Bundle A BES bundle 7° baffle angle; (b) Bundle B - BEU bundle 12° baffle angle.

New seal strip design

From the tests performed with no seal (▲) and with conventional seal strips (■), there were two primary observations:

1. The pressure drop penalty of adding seal strips for Bundle A was significant compared to the relatively modest increase in heat transfer.

2. Shell-side performance as determined through *Pe* and C_{pd} was not improved by the addition of seal strips.

Whilst the conventional sealing arrangement does indeed block the flow in the region between the tubes and the shell, the resulting form drag also disrupts the main helical flow path. Heat transfer enhancement resulting from turbulence generated by the sealing strip is restricted to a local region around the strip. The net result may be that any local heat transfer improvement is offset by a reduced heat transfer rate in the main helical flow.

A number of alternative sealing designs were considered in an attempt to reduce the pressure drop whilst maintaining the heat transfer performance. A new sealing arrangement was developed to reduce the flow of the outer helical stream whilst accounting for its interaction with the main helical flow. The most promising alternative is labelled as "New Seal" in Figures 6 and 7. Figure 6 is a plot of the performance of the new seal on top of the data from Fig. 5. Figure 7 is a plot of *Nu* vs. $Re_{helical}$ for Bundle A. The new seal improved the shell-side performance over the entire range of test conditions for Bundle A. For Bundle B, the tests with new seal strip were carried out using a precision fabricated shell, so the impact of the new seal cannot be easily differentiated from the impact of the new shell. However, it appears that the new seal is likely most effective in the Reynolds number range 100 – 500. For Reynolds numbers below 100 no clear improvement is seen.

Whereas the conventional seal had little impact on the shell-side performance, the new seal actually exhibited a slightly lower pressure drop than the no seal case, and a slightly higher heat transfer coefficient than the conventional seal. This somewhat surprising result can be explained by the fact that the ratio of tube wall to bulk viscosity ratio was relatively high (~6), and by reducing cold bypass flow the new seal enabled a lower average viscosity. At the lowest Reynolds number tested, C_{pd} was increased from under 20% to over 30% and Nu was increase from 7 to 9.

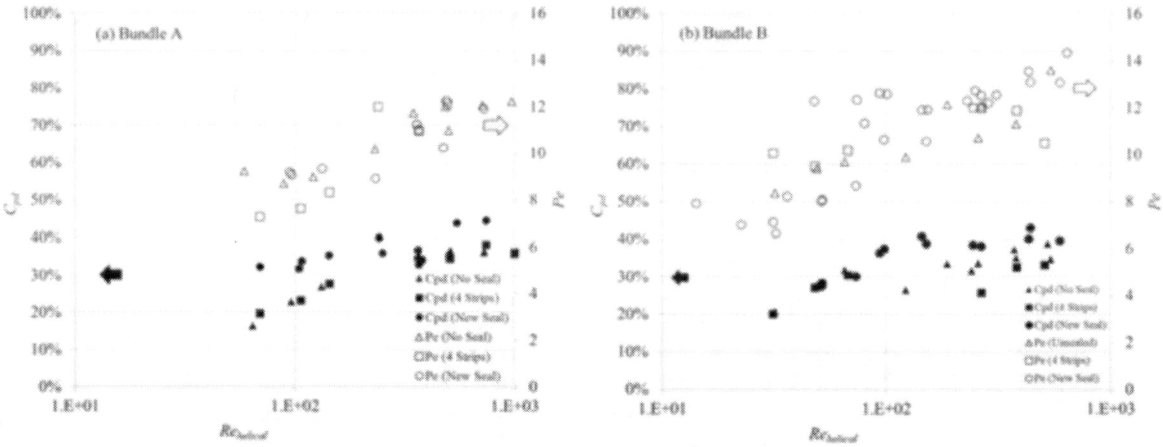

Figure 6 - C_{pd} and Pe as a function of $Re_{helical}$ for Bundle A (BES) with (1) no seal, (2) 4 strips and (3) new seal.

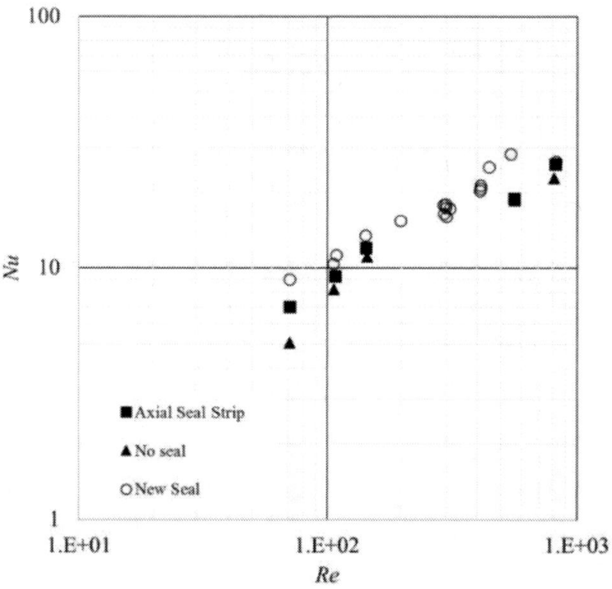

Figure 7 – *Nu* values derived from experimental measurements for Bundle A

Summary

HELIXCHANGER technology has been successfully applied in a number of applications that take advantage of the unique helical flow pattern. The key applications are described along with supporting references and case studies.

For cooling of viscous fluids, HELIXCHANGER heat exchangers can provide significant improvements in heat transfer rates over conventional segmental baffles. In the case study presented an increased cooling rate was measured relative to a conventional cooler with the same surface area. The cooling rates improved 30 - 50% (10 -12 °C) based on geometry changes (HELIXCHANGER bundle retrofit) and 5-10% based on a process modification (diluent addition).

Ongoing research carried out at the HELIXCHANGER test facility in Pasadena, Texas uses a combination of heat transfer, pressure drop and residence time measurements to compare different shell-side configurations. Shell-side performance is evaluated based on the dimensionless parameters C_{pd} and Pe to compare the pressure drop conversion to heat transfer and axial distribution respectively. The combination of these two measurements has proven very useful for characterizing shell-side performance and testing improved configurations. In the study presented in this paper, conventional axial sealing strips were found to provide minimal improvement in overall performance. A new patent pending "helical sealing strip" was tested that was found to provide an improved heat transfer rate at a significantly reduced pressure drop.

Nomenclature

A	Surface area for heat transfer, m^2
C_{pd}	Pressure drop conversion factor, dimensionless
D	Tube outside diameter, m
MTD	Mean temperature difference corrected for flow geometry, °C
Nu	Nusselt number, dimensionless
Pe	Peclet number, dimensionless
Q	Heat duty, kcal/hr
Re	Reynolds number, dimensionless
t	time, s
T	Temperature, °C
U	Overall heat transfer coefficient, W/m^2°C
μ	mean residence time, s
τ	Shear stress, N/m^2
σ^2	Variance

Subscript

B	B-Stream
F	Film
$helical$	Helical flow
PL	Longitudinal flow
w	Wall

References

[1] Master, Bashir I., Krishnan S. Chunangad, Bert Boxma, Graham T. Polley, and Mohamed B. Tolba, "Reduced total life cycle costs using HELIXCHANGER® heat exchangers." In Third International Conference on Compact Heat Exchangers and Enhancement Technology for the Process Industries. Julio, pp. 1-6. 2001.

[2] Master, Bashir I., Krishnan S. Chunangad, and Venkateswaran Pushpanathan. "Fouling mitigation using HELIXCHANGER® heat exchangers." (2003): 317.

[3] Stehlík, Petr, and Vishwas V. Wadekar. "Different strategies to improve industrial heat exchange." Heat Transfer Engineering 23, no. 6 (2002): 36-48.

[4] Stehlik, P., J. Němčanský, Dl Kral, and L. W. Swanson. "Comparison of correction factors for shell-and-tube heat exchangers with segmental or helical baffles." Heat transfer engineering 15, no. 1 (1994): 55-65.

[5] Kral, D., P. Stehlík, H. J. Van Der Ploeg, and Bashir I. Master. "Helical baffles in shell-and-tube heat exchangers, Part I: Experimental verification." Heat transfer engineering 17, no. 1 (1996): 93-101.

[6] Lutcha, J., and Jl Nemcansky. "Performance improvement of tubular heat exchangers by helical baffles." Chemical Engineering Research and Design 68, no. 3 (1990): 263-270.

[7] Master, Bashir I., Krishnan S. Chunangad, and Venkateswaran Pushpanathan. "Heat exchanger." U.S. Patent 6,827,138, issued December 7, 2004.

[8] Barletta, T., White S., and Chunangad K.S., "Designing a crude unit heat exchanger network." www.eptq.com Sour and Heavy Petroleum Technology Quarterly 16 (2012): 3.

[9] Jibb, R. J., M. Brignone, S. Taha, G. Podrebarac, J. Larance, and P. Junge. "HELIXCHANGER® HEAT EXCHANGER-FIELD EXPERIENCE AND ONGOING DEVELOPMENT." Heat Exchanger Fouling and Cleaning XIII 2019, 2-7 June 2017, Józefów (Warsaw), Poland.

[10] Gough, M. & Dröegemüeller, P. & Ellerby, P. & Lee, M. & Master, B.I. & Chunangad, K.S.. (2007). Dual enhanced heat exchangers using Cal Gavin hiTRAN® wire matrix tube inserts along with the ABB-LGHT HELIXCHANGER® technology. 17. 32-41.

[11] Rohsenow, Warren M., James P. Hartnett, and Young I. Cho. Handbook of heat transfer. Vol. 3. New York: McGraw-Hill, 1998:L 17.136 – 17.145.

[12] Jibb R.J., Ebeling C., Taha S., Gallei W., Macedo E., HELIXCHANGER® Heat Exchanger – Experimental and Field Pressure Drop Measurements used to validate CFD model" Heat Exchange Engineering Asia 2014

[13] Roetzel, Wilfried, and Frank Balzereit. "Axial Dispersion in Shell-and-tube Heat Exchangers." Int. J. Therm. Sci. 39 (2000): 1028-038. Print.

[14] Dudukovic, M. P., and R. M. Felder. "Mixing effects in chemical reactors–III. Dispersion model." Reactor Stability, Sensitivity and Mixing Effects (Module E4. 7) (1983).

[15] Roetzel, Wilfried, and Yimin Xuan. "Transient behaviour of shell and tube heat exchangers." Chemical Engineering & Technology: Industrial Chemistry-Plant Equipment-Process Engineering-Biotechnology 16, no. 5 (1993): 296-302.

[16] Roetzel, Wilfried, and Deiying W. Lee. "Effect of baffle/shell leakage flow on heat transfer in shell-and-tube heat exchangers." Experimental Thermal and Fluid Science 8, no. 1 (1994): 10-20.

[17] Palen J. W. and Taborek J., Parametric Response Study of Segmentally Baffled Shell-and-Tube Exchanger Performance, HTRI Rept. S-SS-3-2 (1967).

[18] Tinker, T. "Shell-side Characteristics of Shell-and-tube Heat Exchangers: A Simplified Rating System for Commercial Heat Exchangers." Trans. ASME 80 (1958): 36-52.

[19] Bouhairie S. A., Optimal Application of Shell-side Sealing Devices, HTRI Rept. S-SS-3-28 (2014).

590754 Performance Evaluation Criteria of Enhanced Heat Transfer Technologies Using Swirl Flow

Abstract:

Swirl flow devices are used in industry to enhance heat transfer. The purpose of these devices is to increase performance of an existing heat exchanger, to reduce a new exchanger cost by means of surface area reduction, or to provide a necessary function for the application, such as complete vaporization. At the same time these benefits are realized, swirl flow devices also impact the pressure drop characteristics of the equipment. In order to quantify the benefits of various configurations, select Performance Evaluation Criteria (PEC) are utilized to gauge the impact of several devices by evaluating a wide set of data against constant parameters. This allows the range of effectiveness for the swirl flow devices to be presented.

590181 A Critical Review with Compilation of Pure and Multicomponent Gas Adsorption Equilibria and Isosteric Heat of Adsorption Data

Abstract:

We present a compilation and critical review of the existing literature data on gas adsorption equilibria and isosteric heat of adsorption. The NIST/ARPA-E adsorption database[1] covers wide ranges of pure component adsorption isotherms. However, it does not cover data for mixed-gas adsorption equilibria nor isosteric heat of adsorption. These data are essential for the development of robust multicomponent gas adsorption equilibria model such as adsorption non-random two liquid (aNRTL). The reliable predictions of the models depends on the quality of experimentally measured data and experimental measurement techniques. In this work, we compile and analyze the quality of mixed-gas adsorption experimental data and uncertainty associated with it. We further present prediction of the selected models for several multiple component systems.

References

1. Siderius, D. W.; Shen, V. K.; Johnson III, R. D.; van Zee, R. D., *NIST/ARPA-E Database of Novel and Emerging Adsorbent Materials, National Institute of Standards and Technology.* Gaithersburg, MD, 20899, https://dx.doi.org/10.18434/T43882, (retrieved August 07, 2019).

589847 Simulation of Pressure Swing Adsorbers Using the Moving Bed Model

Abstract:

Simulation of pressure swing adsorption (PSA) units remains a challenge for practicing engineers, which hinders the process intensification potential of the technology. The Moving Bed model, a steady-state model which represents any adsorption process as a series of interconnected continuous adsorption units, has the potential to alleviate some of the difficulties in modeling PSA. [1, 2, 3] In the Moving Bed model, the cycle is divided into multiple constant-pressure operating states bearing similarity to the individual steps of the cycle which allows the properties of the state to be evaluated from equilibrium considerations alone. The operating steps are arranged from highest-to-lowest pressure to represent the driving force gradient exploited during the cycle and then connected by a virtual moving bed. The virtual moving bed does not physically exist in the system but represents the adsorbate and void space hold-up retained in the adsorption column at the end of the operating state. However, modeling a cyclic batch process as a continuous, equilibrium process neglects the rate-based transport phenomena involved. Using the idea of adsorption efficiencies [4] to account for non-idealities like mass transfer resistance and dispersion, similar to the tray efficiencies used in rate-based distillation modeling, the retentate flow rates can be adjusted to match the actual performance of the unit. [2, 3] To showcase the capabilities and current limitations of the model, a case study will be presented which investigates a lab-scale air separation unit. Guidelines for ensuring convergence and robustness will be given and the effect of important parameters will be discussed.

1. Sees, M.D. (2018, October). *Novel Steady-State Process Modeling Methodology for Pressure Swing Adsorption*. Paper presented at the 2018 Annual Meeting of the American Institute of Chemical Engineers, Pittsburgh, PA.
2. Kirkes, T. (2018, October). *Steady-State Modeling of Air Separation by Pressure Swing Adsorption Using a Bed-State Efficiency Model.* Poster presented at the 2018 Annual Meeting of the American Institute of Chemical Engineers, Pittsburgh, PA.
3. Sees, M.D. (2019, April). *Modeling of Gas Separations by Pressure Swing Adsorption Using a Novel Steady-State Methodology.* Paper presented at the 2019 Spring Meeting of the American Institute of Chemical Engineers, New Orleans, LA.
4. Smith, O.J. and A.W. Westerberg. *Eng. Sci.*, 1991, 48(12), 2967 – 2976.

589746 Simulation of Near-Optimal and Valve-Free Operational-Half of Pressure Swing Adsorption Processes for Fair and Facile Evaluation of Process Performance Metrics

Abstract:

In this contribution, we will present a method for simulating *near-optimal* and *valve-free operational-half* of pressure swing adsorption (PSA) processes based on our new modeling and simulation framework, so called the *valve-free model*. The framework automatically determines *near-optimal* operational decisions during dynamic simulation of a PSA process. Also, the model eliminates any necessity for the user to specify inputs related to the detailed operation of a PSA process. Therefore, by using the framework, any non-expert users can easily obtain simulation results corresponding to a *near-optimal* operation of a PSA cycle.

First principle dynamic modeling and simulation of PSA processes has widely been used to capture dynamics of non-equilibrium adsorption phenomena. Furthermore, the approach has potentials to be utilized in many industrially relevant applications such as adsorbent screening, conceptual process design, optimal control, and optimization. Despite these potentials, however, there have been several limitations that prevented widespread adoptions of the first principle modeling and simulation in providing solutions to the aforementioned applications.

One of the major limitations of the current standard dynamic model, i.e. *valve* model, is its mathematically intractable formulation. For instance, in order to mathematically describe an adsorption column, the model needs to be formulated using a coupled system of nonlinear partial differential equations. Typically, solution methods to a such model are quite involved in terms of numerical methods used, conceptually hard to understand, and difficult to implement.

Another major limitation of the *valve* model is related to the user experiences in utilizing the model. Currently, to simulate a *near-optimal* PSA process using the *valve* model, the user has to supply a carefully chosen set of input parameters for a given PSA process simulation. However, choosing a set of input parameters that leads to *near-optimal* process simulation is not so simple as there practically exists so many reasonable ways to specify inputs such that the final simulation results are nowhere *near-optimal*.

Lastly, one of the most critical limitations of the *valve* model is the apparent sensitivity of simulation outputs on user supplied inputs. When designing chemical process systems, having accurate process performance metrics is considered as an essential requirement in determining process feasibility. However, due to the sensitivity issue, performance metrics obtained by the *valve* model are substantially influenced by the user supplied inputs. Furthermore, identifying an optimal set of inputs for *valve* model is computationally exhausting as it requires time consuming *trial-and-error* simulation studies.

Therefore, these limitations prevented a broader group of researchers and practitioners from being able to accurately predict performance metrics related to PSA process. Since enabling chemical industry to reliably assess adsorption based *process intensification* (PI) concepts would lead to a faster and more widespread deployment of such novel technologies, there is a critical need for developing a new dynamic model for simulating *near-optimal* PSA processes.

To address this critical need, a novel *valve-free* model has been developed to enable the user to simulate *near-optimal* PSA processes with *Skarstrom* type PSA cycles. In order to develop the *valve-free* model, the *Skarstrom* PSA cycle was decoupled into two halves including different steps, namely the *operational-half* and *re-generation-half*. Based on the simplified system, *near-optimal* operational policies for the two halves were devised separately.

A *near-optimal* operational policy for the *re-generation-half* was presented in previous talks given during the 2019 Spring AIChE meeting[1] as well as the 2019 Annual AIChE meeting[2]. Therefore, in this oral presentation, we are going to discuss the technical details that are involved in the development of a *near-optimal* operational policy for the *operational-half*.

In order to develop the *valve-free* model, several key assumptions were made to simplify the system and to present a *proof-of-concept* for the proposed modeling framework. The current formulation of the *valve-free* model assumes isothermal, no dispersion effects, binary ideal gas mixture, linear isotherm, and linear driving force adsorption kinetics.

The power of the proposed methodology (i.e. *valve-free* model) lies within enabling the user to run high quality PSA process simulations without having to preconceive the respective optimal operational policies. This enhanced user experience may lead to more wide adoption of process simulation of PSA processes as a tool that can help make critical day-to-day decisions or assessments that are related to operation, control, and optimization of PSA processes.

For the scope of the future research goals, the merge between the decoupled halves will be sought after in hoping to devise an overall *near-optimal* operational policy for PSA processes involving *Skarstrom* type cycles. Sequentially, based on the *near-optimal* operational policy for PSA processes generated by our *valve-free* model, further interrogations on the development of design rules for synthesizing *near-optimal* PSA processes will be made.

Ultimately, by serving as practical tools that can address the identified knowledge gap, *valve-free* model can potentially contribute to the advancement of PI and deployment of various adsorption technology in the chemical process industry.

[1] Kim, T.-H. et al. (2019). Rational Screening of Adsorbents for Natural Gas Upgrading By Pressure Swing Adsorption Using Dynamic Simulation of Process Performance Metrics, April 2, 2019, Spring AIChE Meeting, New Orleans, LA, 2019.

[2] Kim, T.-H. et al. (2019). A 'Valve-Free' Model for Dynamic Simulation of Pressure Swing Adsorption Processes That Automatically Determines Near-Optimal Operational Policies, November 11, 2019, Annual AIChE Meeting, Orlando, FL, 2019.

589775 Data Infrastructure As Foundation to Operational Excellence and Digital Transformation

Abstract:

Is technology a solution looking for a problem? A successful digital transformation builds on the foundations of Operational Excellence. Innovations in the technology we use further these initiatives. The building blocks of these improvements are rooted in the data we collect, compile, and contextualize into operational intelligence. Successful implementations identify desired outcomes and build the data and technical infrastructures to achieve them.

A structured Operational Excellence program begins with clear definitions for:

- What's important?
- Who's responsible for action, who's accountable for success?
- How is success measured?
- How are excursions identified, monitored, and corrected?
- What communications are necessary to ensure improvement?
- What skills and tools are necessary to continuously improve?

Compelling events for transformation may include regulatory compliance, hazard recognition, sustainability, cost controls, quality improvements, and asset reliability as examples. The data required covers the spectrum of real-time operational data, environmental monitoring, quality data, and asset information. A connected infrastructure runs from edge connected IoT devices to the cloud, sharing information across individual assets, plants, enterprises, and communities of connected services.

However, the final transformation does not have to be the first implementation. Like any solution, digital tools evolve as our needs and understanding of the problem change. Our use of the data infrastructure will necessarily evolve by:

- Collecting data into the data historian
- Contextualizing an operational "chart of accounts"
- Integrating with external data sources such as LIMS, ERP, CMMS, etc.
- Formulating streaming real-time analytics, first law calculations, and KPI's
- Detecting and notifying on errors and events
- Leveraging complex analytic engines to develop predictive and prescriptive models

Data should be collected once, as close to the source as possible. The data infrastructure will interface with these source systems. As part of the transformation effort, organizations evaluate missing data for the potential value of the information against the expense of collection. Where possible, organizations can leverage IoT technologies or collect manually.

Demonstrated use cases include monitoring site heat index to prevent heat stress, extrapolating corrosion curves to facilitate mechanical integrity programs, performing both golden batch and multivariate analyses, and collecting asset utilization data for reporting among others. Successful digital transformations provide competitive advantages for these enterprises. We believe people with data can transform our world.

590961 How to Realize Your Digital Transformation

Abstract:

Today, few companies doubt the benefits of Industry 4.0. It is relatively easy to talk about the need for a digital transformation from a high level, but when it comes to planning and executing a specific digital transformation roadmap, it tends to be more challenging. But it doesn't have to be that way. Industry 4.0 has now been around for a few years, which has allowed tried and tested approaches and methodologies to emerge that can be applied to pretty much any size company in any industry.

We will look at the value chain of chemical processing plants identifying application areas for Industry 4.0 technologies such as digital twins. The focus of this industry specific approach is on structured and open data models that break down stand-alone and isolated knowledge silos - in effect integrating the industry relevant value chains. Actionable items include, but is not limited to: the use of multiple digital twins as controlled copies handling business critical information for data-based business management; the continuous assessment of the correct technologies and technology levels to be implemented for each business relevant area; business and site level consulting on industry 4.0 until maturity is reached and digitalization is business as usual in the chemical processing industries

In addition, we will gain insights into how to shape your own future by combining novel and mature technologies in new and exciting ways, which includes project prioritization. It is a critical exercise for companies in formulating a digital transformation because it enables them to identify areas that will generate the greatest value and evaluate each of them based on specific criteria. These can then be prioritized, which will help decision-making and effective resource allocation. This may sound like a simple step, but it requires dialog, involvement from several stakeholders and a framework to put it all together. Industry 4.0 is *not* a one-size-fits-all approach.

By attending this presentation, you will gain the following insights:

- Gain high-level insight into the concept of 4.0 technologies applied in chemical processing industries.
- Gain site-level insight into digitalization use cases transforming our industry now!
- Basic understanding of how to identify Industry 4.0 projects.

View of how to best prioritize them in terms of resources, ROI, digital maturity level and more.

589365 Piloting Digitalization at Dupont

Abstract:

Okay, you've decided that it's critically important to begin going down the digitalization path. Now what? This presentation will cover DuPont's efforts to date to implement pilot digitalization projects making use of data from existing assets and provide some learnings that you may benefit from in your own piloting efforts.

Specific questions that will be answered and topics that will be covered include...

- Implementation in a production facility making use of real world data
- Selection of digitalization options to be piloted
- Who was involved in the implementation process?
- How long did it take to complete the piloting?
- What support did you receive during the piloting?
- What specific digitalization tools/apps were piloted?
- What was the outcome of the pilot testing?
- What would DuPont have done differently?
- What are the next steps?

583338 Keynote: Industrial Cleaning: towards condition-based cleaning

Abstract:

Cleaning of industrial equipment creates risks, both for the persons who perform the cleaning activity and for the equipment that is being cleaned. Often specific knowledge to failure modus are unknown by the persons performing the cleaning and too often heat exchangers are being damaged due to the lack of knowledge and experience of the people making the cleaning decisions. This keynote addresses some basic guidelines towards the following encountered failure modes: e.g. corrosion, erosion, overheating, incompatible cleaning chemicals. Furthermore personal safety items will be discussed.

589781 Condition-Based Cleaning Models for Industrial Heat Exchangers

Abstract:

Periodic cleaning of industrial equipment is often necessary to maintain plant performance over long operating periods. Although cleaning decisions (*i.e.* which heat exchanger to clean, when to clean it and – crucially – how to clean it) should be based on economic drivers (including trade-offs between operating costs, throughput of the installation and quality of the cleaning preformed), they are very often made based on operators' experience and simplistic – often misleading – performance indicators (e.g. fouling resistance, pressure drops). Moreover, operators may not be aware of which cleaning technologies are available to them and which is the optimal option to use .

In this paper, a framework integrating condition-based cleaning models that capture kinetics and efficiency of a cleaning action as a function of the characteristics of fouling deposits (e.g. composition, layering, degree of coking) is presented. The approach is used to provide actionable information on the best method to be used, duration of the cleaning and expected economic benefits, depending on actual equipment conditions. Results are presented for individual heat exchangers and entire networks.

588894 Monitoring-Based Fouling Mitigation in Crude Unit Preheat Trains Using Lab and Field Fouling Units

Abstract:

The physical and chemical mechanisms associated with crude-oil fouling are complex, which hinder effective applications of mitigation methods for the preheat-train (PHT) exchangers. In order to develop an effective mitigation strategy, it is essential to perform a root-cause analysis (RCA) to identify the source(s) of fouling precursors. The next step is to evaluate ways to interrupt or eliminate the source of fouling precursors and conditions under which they are generated.

The multi-year investigation in this presentation focused on monitoring-based mitigation to systematically perform the root-cause analysis using lab and field fouling units. The analysis included the effects of crude-oil blending, physical parameters (such as flow velocity and wall shear stress), wall temperature (heat flux), performance of desalter, iron-sulfide induced fouling, asphaltene conversion and the effects of inorganic salts. The fouling data obtained with lab and field-fouling units were used to develop the threshold fouling model, which was then used to develop maps that help identify PHT exchangers that are in the high-fouling zone. The threshold models were also implemented in Hexxcell Studio™ to assess the economic impact of fouling and predict future performance in the pre-heat train operation.

The combination of the fouling data from the lab and field fouling units and the simulations were used for the root-cause analysis based on which an effective mitigation strategy can be developed and its benefits assessed.

590340 Wash Oil Control:Inferred Versus Measured Rates

Abstract:

Crude oil distillation towers use a wash section to remove entrainment from the feed. This section requires wash (wetting) liquid to maximize performance and prevent the wash section from coking. This is a difficult control problem. The liquid rate has a sharp optimum. Too low a rate, and the wash section cokes, shutting down the unit. Too high a rate and significant yield or capacity losses can result. Being wrong in either direction has a high cost.

The liquid rate can be controlled by either direct measurement and control or by inferring a rate based on energy and mass balance around the unit and controlling a secondary stream. The direct measurement system has higher capital and operating costs. The indirect method has lower capital and operating costs, but has a higher risk of upsets and larger uncertainties.

This presentation reviews the results of Monte Carlo modeling of a crude unit to evaluate the economic benefits of the direct approach versus the uncertainties in the indirect approach.

591013 Digital Risk Management and Process Safety

Abstract:

Most organizations recognize IEC 61511/ISA-84 as the global best practice for functional safety management. The challenge is in finding the most effective way to fully comply with these standards over the operation and maintenance phases, which comprise 99% of the asset lifecycle. Traditionally, the front-end engineering work from risk assessment to SIS design has been document-based using one-seat calculators and other one-off, disparate solutions that make it nearly impossible to leverage while the plant is running.

The Layer of Protection Analysis (LOPA) represents the foundation of a functional safety management program and this can assist organizations towards achieving their compliance goals. This forms part of a Digital Risk Management solution which identifies mismatches between operating data and assumptions during design in order to eliminate unnecessary risk. It also leverages internal work processes rather than implementing new ones in order to optimize efficiencies. Implementation of such a solution avoids any culture change or the constant add-ons which are typically involved in lifecycle approaches simply to monitor Tier 3 and Tier 4 metrics as identified in API RP-754.

This presentation will help organizations understand why a Digital Risk Management solution is required in order to utilize the entire process safety investment made by an organization. It will also explain how templates standardize PHA/LOPA reports in order to minimize additional training and configuration costs, as well as how absorbing PHA/LOPA files into the Process Safety Monitoring environment will best assist organizations to focus on the appropriate severities.

591012 Control System Architecture Selection in the Age of the Iot

Abstract:

Selecting a control system today presents many challenges to prospective users. The three main categories for a basic process control system (BPCS) are the distributed control system (DCS), programmable logic controller (PLC), and supervisory control and data acquisition (SCADA) system. A user must additionally decide whether a safety instrumented system (SIS) is required and, if so, how that system interfaces or interacts with the BPCS.

Meanwhile, we are in an unprecedented age of rapidly advancing technologies. They present a quandary to a prospective control system user. In light of the new technologies, will today's BPCS architectures quickly find themselves obsolete? Or are there ways to evolve the control system to work with analytics, artificial intelligence, digital twins, machine learning, and "Internet of things" (IoT) architectures that comprise a new generation of sensors, wireless networks and cloud-based services?

Furthermore, will the BPCS support the digital transformation that your company has been planning and could it evolve into an open system?

This presentation will review today's BPCS and safety technologies, describe the emerging "IoT" technologies and provide a roadmap to make the best of the latter while additionally evolving to support open architecture initiatives such as the Open Process Automation Forum (OPAF) O-PAS standard.

590557 Supporting Digitalization Strategies Around Turbomachinery â€“ Security First

Abstract:

Industrial Control System (ICS) Cybersecurity is a major topic of concern in the energy industry today. This is primarily driven by the two imminent needs: identifying and minimizing current risk exposures – the seriousness of which has been emphasized by the recent attacks against safety instrumented systems – and paving the way for achieving higher level of productivity in the digital age.

As the businesses strive towards more integration between process operations and enterprise to benefit from knowledge centralization, data mobility, artificial intelligence, etc., cybersecurity capabilities of ICS vendors must also mature to be able to support customers' digitalization strategies.

In this presentation we will report on CCC's cybersecurity journey as we utilize ISA/IEC 62443 series of standards in implementing Security Development Lifecycle (SDL) into the organization – from design, delivery, to lifecycle support. Methods adopted through its product development process are outlined – including vulnerability identification testing and communication robustness testing – as well as product and service developments for defense-in-depth.

As a part of CCC's digitalization solutions, a series of new product capabilities will also be demonstrated – including model-based input validation, turbomachinery performance monitoring, and secure remote access to high-quality turbomachinery control data.

590101 Intelligent Operations: A Bold New Approach to Achieve Operational Excellence and Enhance Profitability

Abstract:

Industry continues to seek new ways to enhance financial performance. Intelligent Operations is a bold new approach to achieve Operational Excellence (OpEx). It uses digital transformation to optimize production, minimize equipment downtime, enhance human performance, and manage operational risks. To do this, companies need to break down existing departmental, management, cultural, and data management silos. This presentation will examine the key aspects of Intelligent Operations and explore an approach for implementing a strategy that delivers on the promise of direct data from equipment, real time field inputs from a mobilized workforce, and the visibility possible into fleet-wide operations from an integrated view of operations, safety, compliance and workforce.

589578 SDA-Enabled Solutions for Unlocking IMO2020 Constraints

Abstract:

Converting residues are top priorities for refineries, globally, in the upcoming IMO 2020 Marine Bunker fuel regulation environment. Generally, refineries use either hydrogen addition or carbon rejection technologies for converting residues into high value products. Long residues or vacuum tower bottoms (VR or VTB) with limited amounts of metals and Conradson carbon can be hydrotreated; however, hydrotreater products require further treatment for making final products. The residue hydrotreaters require high pressure and frequent changes of catalyst, making them both capital intensive and high in operating costs. Short catalyst cycles also reduce on-stream factor.

Short residues are usually converted using thermal processes such as thermal cracking and delayed coking. Nonetheless, all products need to be further processed to achieve transportation fuel quality products, adding to the cost of the resid upgrading schemes.

ROSE is the market leading Solvent deasphalting (SDA) technology that can be relatively easily integrated with FCC and/or hydrocracking (mild-to-high pressure, fixed bed as well as ebullated bed hydrocracking) and has been commercially proven and relied upon to meet the need for higher yields of transportation fuels while improving the quality to meet new gasoline and diesel specifications at comparable operating costs and complexity of operations. Combining high-lift ROSE with high conversion hydrocracking and an integrated solidification system would mean a significant reduction in fuel oil exposure and result in higher overall margin than other resid upgrading options described above.

We will present schemes using combinations of solvent deasphalting, hydrocracking and solidification processes. In our presentation we will discuss economic and operating benefits of other bottom-of-the barrel schemes and commercial examples (i.e. debottleneck VDU and/or synergize with cokers), where applicable.

589724 IMO 2020: Opportunities for Refining Margin Enhancement

Abstract:

IMO (International Maritime Organization) 2020 slashes marine sector emissions in international waters. The sulfur emissions from marine fuels will have to be reduced by 80% and a large part of this will achieved by switching to low sulfur fuel oils. IMO 2020 compliant fuel oils will have a maximum sulfur content of 0.5wt% as opposed to 3.5wt% prevalent today. This will be a large change considering that 3.9 Million Barrels/day of Marine fuels was consumed in 2017.

The market response to the change is still developing, price of IMO 2020 compliant fuels are inching towards ULSD and the price of Heavy Sulfur Fuel oils (HSFO) has dropped by about 15 – 20 USD/barrel and is trading at about half the price of Brent Crude Oil. This scenario is likely to continue.

This paper investigates the refining strategies that can upgrade these attractively priced HSFO streams in existing refineries to high value products by modifying the catalyst systems. This kind of approach does not require big ticket capital expenditure but will significantly improve the refining margins.

615159 Trends in Water & Wastewater Treatment: Challenges, Innovation and Progress: Plenary Session

Abstract:

Water Reuse in Circular Economy: From Not Potable to Potable - Municipal and Industrial- Val Frenkel

Currently we facing significant increase in water reuse needs due to the several reasons such as, but not limited to climate changes, increase of natural disasters and uncertainty in water supply, lack of water sources, and increase cost of water, tight discharge criteria, regulatory pressure, and corporate mandates. Each project development needs right approach and right strategy.

This presentation will show evolution from not potable to potable water reuse, both Indirect Potable Reuse (IPR) and Direct Potable Reuse (DPR), best practices in water reuse as well as the leading utilities and industries in water reuse. The approach to implement economical water reuse will be presented, how to select the treatment process and several successful water reuse case studies based on the presented approach will be shared.

New Trends in Recovery of Nutrients from Point and Distributed Sources - James Barnard

The use of biological systems for removing nitrogen and phosphorus from used water. These systems are used in some form or another universally. Nitrogen is mostly returned to the atmosphere, but phosphorus can be recovered as struvite or some other re-usable product. The latest research concerns a better understanding of biological phosphorus removal to make the process more reliable to cope with varying wastewater characteristics as influenced by seasonal temperature and rainfall. The present quest is for cost-effective retrofitting of existing plants to remove phosphorus to very low levels. However, achieving greater efficiency at point sources does not solve the major problems of plant nutrients from agricultural run-off, reaching lakes and estuaries. Efforts are made to grow and harvest algae as a means of removing these plant nutrients from receiving water which could result in energy recovery.

Risk-Based Assessment of Refinery and Petrochemical Plant Effluents - Charles Meyer

This presentation will discuss assessment of hazard and risk of effluent discharges from refinery and petrochemical plants in receiving waters. Besides monitoring of conventional (e.g. pH, BOD) wastewater parameters and specific chemical constituents, use of Whole Effluent Toxicity (WET) testing is used extensively in North America to provide a holistic hazard assessment of effluent discharges. Combining this with information on exposure (e.g. dilution, receiving water characteristics) can facilitate an overall risk assessment. Using a tiered approach to risk-based assessment can help to achieve risk to the environment "As Low as Reasonably Practicable" (ALARP) in a way that minimizes unnecessary testing. Use of Toxicity Identification Evaluation (TIE) to help understand causes of toxicity can be used to then develop risk reduction strategies.

Partnership between Rice and City of Houston to Advance Sustainable and Resilient Water Systems - Lauren Stadler

This talk will discuss ongoing collaborative projects between Rice University and the City of Houston related to advancing sustainable and resilient water management. Projects include quantitative modeling to assess the economic and environmental impacts of distributed direct potable reuse; modeling and pilot-scale studies on wastewater treatment plant resiliency to wet weather events; and wastewater monitoring for tracking community infection dynamics. We will highlight the benefits of university-utility collaborations for conducting both applied and fundamental research, and for developing site-specific solutions for water management systems and protecting public health.

Intensifying Treatment and Resource Recovery SWIFTly and Pragmatically at HRSD - Charles Bott

This talk will cover topics ranging from sidestream deammonification implementation at HRSD, recent advancements in mainstream partial denitrification anammox, and the new HRSD indirect potable reuse and aquifer recharge program known as SWIFT. HRSD's Sustainable Water Initiative for Tomorrow (SWIFT) is a water purification project in eastern Virginia that will further protect the region's environment, enhance the sustainability of the region's long-term groundwater supply and help address environmental pressures such as sea level rise and saltwater intrusion. The project will take HRSD's already highly-treated water that would otherwise be discharged into the Elizabeth, James or York rivers and implement advanced water treatment to produce drinking-quality water. The finished water will then be injected into the Potomac aquifer, the primary source of groundwater throughout eastern Virginia. This could help improve water quality in local rivers and the Chesapeake Bay, while ensuring a sustainable source of groundwater for future generations. In partnership with Virginia Tech and others, HRSD has been piloting potable reuse advanced treatment systems, initially comparing the removal of TOC, pathogens, emerging contaminants and disinfection byproducts in two advanced treatment trains: MF/RO/UVAOP versus Coag/Floc/Sed/Ozone/BAF/GAC/UVD. We continue to perform pilot testing work on the ozone/BAF-based system to address a wide range of research objectives. In addition, HRSD is now operating the SWIFT Research Center (SWIFTRC). This facility includes a 1.0 MGD demonstration-scale advanced treatment system using the ozone/biofiltration-based configuration, a large research laboratory, a flexible and adaptable space to house the advanced treatment pilot systems and soil columns, a recharge well and network of monitoring wells, an extensometer station to measure land subsidence/rebound, and a SWIFT public outreach center.

The Wastewater Industry- Current Trends and Future Direction - R. Benson Pair

This presentation draws on the knowledge and experience of experts in wastewater treatment to offer their insight into the current trends and future direction of the wastewater treatment industry. The wastewater treatment industry is a mature industry, but continues to face technical, regulatory and operational challenges. The insight and perspective of individuals from several facets of the industry has been solicited, including academia, professional societies, consultants, technology and equipment providers, regulators, owner/operators, and design engineers.

591159 When Disasters Happen: A Collaborative Approach Towards the Community's Recovery Process

Abstract:

The 2020 Virtual AIChE Spring Meeting & 16th GCPS Wednesday keynote "When Disasters Happen: A Collaborative Approach Towards the Community's Recovery Process" will be presented by Thomas Munoz, Emergency Manager and Homeland Security Director with Texas City. As the Emergency Manager and Homeland Security Director with Texas City, the home of 19 chemical plants, he is responsible for the safety of all its citizens along with managing the crisis communication plan with local industry and major stakeholders.

Safety and disaster planning is vital to the recovery process after a major incident. As key critical infrastructure industries, issues may arise that have not been planned for in advance, such as the recovery from an incident. Developing and maintaining partnerships with local partners in your community will support addressing unforeseen challenges. Recovery efforts include planning with key partners and understanding the phases of recovery, which also consider the mental health and well of everyone involved in the response, as well as the community impacted. In this keynote address a discussion of pre-incident planning considerations and strategies as well as lessons learned/after action improvements post-recovery will be discussed. Additionally, collaboration considerations regarding local resources to promote recovery in the event of a workplace incident will be discussed.

C0-presenter Amanda Groller, MS.Ed., LPC will discuss the mental health impacts of a disaster.

591582 Plant of the Future

Abstract:

For the "Plant of the Future", there are a lot of promising options. Some of the future technology ranges from automated start-up, operations and shut-downs to Artificial Intelligence (AI) predicting process outputs and machine failures to remote locations being fully controlled due to internet connected sites. Some future changes could be process technological changes, such as electrification and closed loop economies. Another key point that will have to be addressed is ensuring site security and safety: both physical and electronic.

In this panel we will talk about opportunities and early successes and pitfalls and challenges. Topics will cover greenfield builds and brownfield expansions and retrofits.

589797 Key Note: Heat Exchanger Networks 40 Years after the Discovery of the Pinch

Abstract:

Pinch Analysis (PA) has provided a lot of insight related to heat recovery in complex industrial processes and total sites. A fundamental decomposition at the heat recovery Pinch provides important design guidelines for heat exchanger networks; both in the grassroot case and for retrofitting existing plants. The Pinch decomposes the process into a heat deficit region above the Pinch and a heat surplus region below the Pinch. This thermodynamic insight has implications far beyond heat exchanger networks and has been formulated through the Appropriate Placement concept for grassroots and the Plus/Minus principle for retrofits. Applications include the integration of chemical reactors, distillation columns, evaporators, heat engines and heat pumps. Important graphical diagrams such as the Composite Curves and the Grand Composite Curve have been developed. Manual design based on PA is, however, time consuming and extremely difficult for industrial size problems due to the multiple economic trade-offs involved.

Optimization formulations offer tools for handling these trade-offs in a rigorous way, and they also provide a possible framework for automatic design. Both deterministic approaches (e.g. Mathematical Programming, MP) and stochastic search algorithms have been proposed to address the Heat Exchanger Network Synthesis (HENS) problem. The HENS problem involves both (i) discrete decisions such as matching of streams and sequence of units, and (ii) nonlinear relations in process models and cost equations. As a result, the problem is a so-called Mixed Integer Non-Linear Programming (MINLP) problem. The computational complexity increases exponentially with the size of the problem (e.g. the number of streams and utilities), and the non-convex nature of the non-linear relations causes gradient-based optimization algorithms to end up in a local rather than the global optimum. Stochastic search algorithms trying to mimic processes in nature (e.g. Simulated Annealing, Genetic Algorithms, Particle Swarm, etc.) try to overcome the problems related to local optima. Experiences show, however, that very long computing times are required, and there is no guarantee for global optimum. Nevertheless, good (near optimal) solutions are often obtained in reasonable times.

In order to take advantage of the best of the two schools of methods (PA and MP) while avoiding their deficiencies, hybrid methods have been developed. One such approach, SeqHENS, will be described and used in this presentation to solve an industrial size Heat Exchanger Network Synthesis problem. As the name indicates, the method is sequential in nature, and the large and complex MINLP has been broken down into more easily solved problems in the form of LP, MILP and NLP. Maximum insight from PA has been utilized to design a framework where the designer (engineer) can explore various loops where the level of heat recovery, the number of units, and the matching between hot and cold streams are varied in a systematic and efficient way.

Unfortunately, both schools of methods (PA and MP) have used rather simplified models for heat exchangers and utilities. In addition, the cost models used are very simple, and a number of issues that are important in industry have been neglected in these models. Examples of such issues are stream splitting, piping, layout, space, weight, materials of construction, pressure ratings, flow configurations, internal details for heat exchangers, as well as issues of process operation and control.

Another problem with the above-mentioned methodologies for heat recovery in industrial processes is the fact that only thermal energy (heating/cooling) is considered with temperature as the quality parameter. Mechanical energy (work and power) is not considered, and the same applies to the pressure of process streams. An important extension of HENS has been emerging over the last 10 years referred to as Work and Heat Integration, where thermodynamic principles are used to design Work and Heat Exchange Networks (WHENs) consisting of heat exchange equipment (heat exchangers, heaters and coolers) and pressure change equipment (compressors, pumps, expanders and valves) in order to improve energy efficiency while considering both heat and work. The main idea is to include heat from compression and cooling from expansion in the classical heat recovery problem. In this framework, process streams can temporarily be used as utilities or as working fluids in heat pumps and refrigeration cycles. Both PA and MP based approaches have been proposed to solve WHEN problems.

The presentation will focus on the high-level topics related to the development and applicability of the methods for heat exchanger networks (and simultaneous work and heat integration) in an industrial real-life setting, while avoiding details on mathematics and algorithms. An industrial-size problem will be used to illustrate recent developments.

589002 Heat Exchanger Troubleshooting - Root Cause Analysis Reveals Shortcoming of Standard Design Methodologies

Abstract:

Heat exchangers are remarkably complex pieces of equipment, with most units encompassing a combination of turbulent, two-phase flows with boiling/condensation occurring in convoluted geometries. All Design and Rating methods include significant simplifications to facilitate performance prediction of these industry work-horses - including assumptions of pure cross or longitudinal flow and well-mixed liquid-vapor phases. In this talk, we will go over two real industry case studies.

The first one involves multiple steam generator operating in parallel suffering from repeated tube failures. These steam generators are thermosiphon reboiler units connected to a common steam drum, and each exchanger is a TEMA AES type with vertical cut double segmental baffles. A visual inspection of the extracted bundle and shell interior reveal erosion damage on the outside surface of tubes in the upper rows and at the shell outlet nozzle connection. The failure led to multiple plant shutdown periods, leading to significant loss of throughput and revenue. During root case analysis, the following topics will be discussed: piping flow loop calculations around the heat exchanger, tube vibration, erosion/ corrosion prediction guidelines and phase separation. The heat exchanger rating software *X*ist will be utilized in this analysis, and hand calculations methods will be discussed. Insights gained from research performed at Heat Transfer Research Inc. will also be discussed, including high speed videos of phase separation inside horizontal shell-and-tube heat exchangers.

The second case study involves a vacuum overhead condenser, where de-superheating, condensation and subcooling are all expected to occur in the same unit. Such a design has multiple vulnerabilities due to uncertainty in design methods for desuperheating and subcooling, as well transition between different mechanisms. In particular, the use of a single bulk temperature profile can lead to divergence between prediction and performance when significant flooding is expected in the condenser. This unit has never provided the required output, despite being significantly over surfaced. During troubleshooting of this unit, various fundamental condenser design issues will be discussed: removal or venting of non-condensables, condensate drainage and removal of tubes. The effect of condenser design on its performance, with respect to non-idealities like vapor bypass will also be discussed.

Hopefully, the talk will impress upon the attendees the importance of sound engineering tools and engineering judgement in designing heat exchangers, and how research can help simplify some of the mystery often associated with heat exchanger troubleshooting.

589492 Simulation-Based Digital Twins for Heat Exchanger Operation and Maintenance Management

Abstract:

The insight of simulation-based simulation in conjunction with advancement in collection and usage of sensor data enabled by the wireless technology has created great opportunities. It can drastically reduce risks associated with cost and schedule overruns resulted in less warranty cost and unexpected downtime. Simulation technology plays a central role in increasing the ROI by providing insight from the deployment stage of the IoT platform to its operation and end of life. This includes sensor placement optimization to physics-based data interpretations and physics-based decision making. Additionally, it is being used for monitoring, diagnostics, prognostics, and prescriptive resolutions to optimize asset performance and utilization and record the digital footprint of the system. This talk will showcase the implementation of simulation based digital twin approach for a shell & tube heat exchanger. The power of traditional time consuming and complex simulations will be used to perform real-time prescriptive and predictive analytics of a heat exchanger. This approach is generic and can be expanded to other assets in the field as well.

589455 How New Modeling Capabilities Enable Digital Transformation to Improve Refinery Performance

Abstract:

This paper will address refinery modeling problems that have presented difficult challenges to simulators, including properly characterizing assays, estimating properties, predicting behavior, and improving planning. This paper focuses on advances in modeling such as molecular modeling-based properties estimation that allow for quicker and more accurate refinery models. These models can evolve into digital twins of the refinery asset, leading to improved safety, operability, and revenue. Examples of successful applications outlining the technical challenge and how it was overcome will be shared along with the estimated economic impact that these improved models have delivered.

Integrating a proprietary simulation tool into a commercial process simulator

Ralf Notz[1], Rafael Rocha[3], Aubrey Kelm[3], Kailee Fujimoto[2], Torsten Katz[1], Jens Schwärzli[1], Agnes Dittel[1], Myrian Schenk[4]

[1] BASF SE, Ludwigshafen, Germany

[2] BASF Corporation, Houston, Texas, United States of America

[3] KBR Inc., Houston, Texas, United States of America

[4] KBR, Leatherhead, United Kingdom

1. INTRODUCTION

The design portion of an engineering project typically goes through several stages. The major stages are Conceptual Engineering, Basic Engineering (normally known as FEED) and Detailed Engineering, or EPC. Once the process concept for the project has been selected (normally during Conceptual Engineering), the Basic Engineering phase begins with the generation of Heat and Material Balances (H&MBs) of all process unit operations. These H&MBs will become the basis for the equipment sizing, line sizing, data sheets, and several other deliverables.

The H&MBs are generated using commercial process modeling environments (PMEs). The corresponding simulations are performed by the Engineering Contractor. For "specialized" unit operations or processes provided by process licensors, often in-house or proprietary simulation tools are employed.

Normally this simulation task is left to the Licensors, i.e. the companies that provide technology licenses for that particular process. Contractors and Licensors need to work together to get the full H&MBs for the project. In addition to the logistics and schedule issues, the workflow between the Licensor tools and the PME is typically inefficient and cumbersome. Integrating these simulation tasks can take very long time, but recently with the advances of digitalization and applying CAPE-OPEN interfaces to embed Licensor tools into PMEs this is starting to change.

This paper describes how CAPE-OPEN compliant Licensor tools can be used in CAPE-OPEN compliant PMEs. This allows to bring the necessary information from the Licensor's process simulation into one location (the PME) for creating Contractor's project deliverables. The result is a streamlined workflow which is less prone to transcription errors and saves on manual effort.

2. DIGITALIZATION

A prerequisite for a successful project is effective communication of all parties involved, namely: Clients, Contractors, Licensors and Vendors. A key component in this communication, is gathering and maintaining all the required information on a central, easily accessible location, allowing for the seamless flow of information between them. This task is made easier by an integrated software solution which can act as a data repository, quality check, and fulfill the requirement of distribution of all information.

Digitalization has brought the possibility of automating the activities in partial or complete form. This has led to reduction of manual effort, and thereby the time it takes to institute changes and the risk of transcription errors. The process simulation is a key component of this work process and many companies have streamlined the process of generating deliverables from PME data; i.e. the required design documents can be generated automatically once the simulation has been approved.

With the introduction of proprietary simulation software from the Licensors into the Contractor simulation, this process is further automated. Historically, data from the PME had to be manually transferred to the proprietary simulation software (or sent to the Licensor) then, the results had to be manually transferred back to the PME and onto deliverables, sometimes taking several iterations. This process of manual data entry is inefficient and incompatible with the digitalized work process that is required these days in the engineering world.

3. CAPE-OPEN

The described limitation in data transfer between a PME and a proprietary simulation software for a digitalized workflow can be overcome by CAPE-OPEN. The CAPE-OPEN standard was established in the end of the 1990s and defines rules and interfaces that allow CAPE (Computer-Aided Process Engineering) applications to interoperate. For maintenance, for support of the existing specifications and for further development the not-for-profit organization CO-LaN (the CAPE-OPEN Laboratories Network, www.colan.org) has been established in 2001. CO-LaN also facilitates the implementation of CAPE-OPEN interfaces through software tools and services.

The CAPE-OPEN standard allows the communication between a PME and a process modeling component (PMC), which in general is intended to carry out a well-defined function with limited scope. This could be e.g. the calculation of physical properties, the calculation of a particular unit operation or the numerical solution of certain types of mathematical problems. Also, a proprietary simulation software can be considered as PMC. Thus, the CAPE-OPEN interface can be used to embed a CAPE-OPEN compliant proprietary simulation software as PMC into a CAPE-OPEN compliant PME. Material streams, energy streams and parameters can be connected to corresponding inlet and outlet ports at the CAPE-OPEN interface. Thus, running the process simulation in the PME, this will automatically transfer data to the inlet ports of the PMC, trigger its calculation and return the results via the outlet ports to the PME. These results are then available for further use in the PME.

4. OASE®

In this paper, the integration of BASF's proprietary simulation tool OASE connect into an Aspen HYSYS® flowsheet of KBR via CAPE-OPEN interface is shown for the removal of acid gas (CO_2 and H_2S).

Acid gas has to be removed from gas streams to meet the specifications of different production plants. This unit operation is normally required in natural gas processing plants (including LNG production), ammonia plants and/ or synthesis gas plants (syngas, methanol plants) to name a few.

Under the OASE® brand, BASF provides customized gas treatment technologies for many application fields. The OASE® process for the removal of acid gas for a natural gas plant uses an activated amine solution, which absorbs CO_2 and H_2S from the feed stream. The rich amine is regenerated in a stripper unit for re-use in the absorber and the acid gas stream is sent for disposal or reinjection.

An exemplary flow scheme of an OASE® single-stage absorber process configuration is shown in Figure 1.

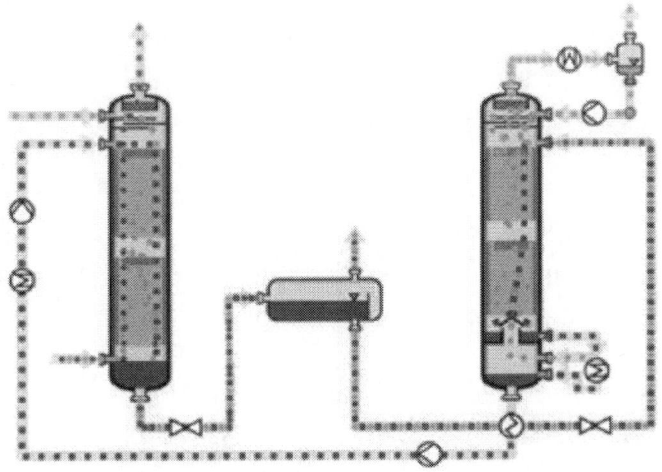

Figure 1: Exemplary flow scheme of an OASE® process configuration for a single-stage absorber with HP flash and stripper regeneration

Design and rating simulations for the OASE® acid gas removal unit (AGRU) can be calculated with the proprietary simulation tool OASE connect, which is provided to engineering contractors and to customers as a standalone server client application. The underlying model is a rigorous process model based on MERSHQ equations (Material balances, Energy balances, mass and heat-transfer Rate equations, Summation equations, Hydraulic equations, eQuilibrium equations) and includes BASF's OASE® specific thermodynamic data.

For an AGRU, as shown in Figure 1, the following battery limits streams exist:

- feed gas stream - entering at the bottom section of the absorber, is separated into
- sweetened treated gas stream- leaving the top of the absorber;
- acid off gas stream- leaving the top of the regenerator;
- flash gas stream- leaving the HP flash;
- water make-up stream (or a bleed stream depending on the overall water mass balance) - entering at the top of the absorber;
- steam (or oil) for the reboiler- entering and leaving the reboiler at the bottom of the stripper
- water (or air) to the condenser - entering and leaving the condenser at the top of the stripper;

- water (or air) for the lean solution cooler, which cools the amine feed to the top of the absorber, entering and leaving the cooler.

Internal streams of the aqueous OASE® solution determine the absorption process, but do not form part of the AGRU boundary limits, between the Licensor and Contractor- however, these are also required for the automation of the work process of an engineering project.

For the described battery limits streams to/from the AGRU as well as for internal streams of the AGRU, manual data transfer between the standalone OASE connect tool and the PME is required when running process simulations. Usually several iterations are required in a PME until the simulation is converged, which contributes to the complexity of the work process. It's also important to note that, the stream conditions, i.e. mass and volumetric flow, temperature, pressure and vapor fraction together with the composition (normally more than 50 components), enthalpy and molecular weight are, at a minimum required to fill in the H&MB and subsequent documents in the design process. The number of H&MB and rating cases required to fully describe the process and to ensure the design will perform as required, add to this complexity. The manual 'copy&paste' of all this information has been the normal practice till now.

BASF has been working to enable the integration of the in-house simulation tool OASE connect into a PME. As such, a CAPE-OPEN interface has been implemented into this simulation tool according to the CAPE-OPEN standard. The implementation project has been split into three phases:
1) specification,
2) implementation and
3) testing.

In the first phase, the required battery limit streams need to be defined for later implementation of material ports and energy ports. Furthermore, ports for parameters as well as options for component mapping needed to be defined. Additional specifications were required e.g. for the visualization of the simulation progress, providing simulation messages or implementation of the authentication procedure.

In the second phase, beside the implementation of the CAPE-OPEN interface itself, OASE connect as PMC also needed to be modified to exchange the required data with the CAPE-OPEN interface.

In the third phase, two parts of testing have to be distinguished. First comprehensive internal tests have been carried out in several PMEs to assure the functionality of the CAPE-OPEN interface. Second, the product "OASE connect with CAPE-OPEN interface" is tested by Engineering Contractors.

5. EXAMPLE

Testing of the OASE connect CAPE-OPEN interface was undertaken by KBR in the fall and winter of 2019. The interface was utilized in a simulation of an LNG plant. The overall flow diagram for this plant is depicted in Figure 2. The unit to be modeled by the interface (the acid gas removal – amine system) is highlighted in purple. The block flow diagram only shows 5 streams passing between this unit and the rest of the plant, constituting the battery limit. The utility streams are not shown here, but they are also part of the battery limit and need to be connected to the rest of the plant.

In the process of developing/designing the project, the ideal configuration of the acid gas system first had to be determined. This decision-making process takes into consideration the specifications of the sweet gas stream as well as the equipment within the unit and how the acid gas stream (also known as waste streams) affect equipment outside of the unit (e.g. if the acid gas is routed to a thermal oxidizer unit for disposal or sent to a compression/dehydration unit for injection).

In utilizing the CAPE-OPEN interface, within the PME multiple configurations could be tested quickly, so that all the varying factors could be addressed, and more importantly without manual transcription of data between different simulation tools. This saved time and effort for engineers and increased confidence that the selected solution was optimum, as many other possibilities can be thoroughly explored.

As simulations of additional cases for the LNG plant design were developed, there was no lag time in updating the AGRU unit utilizing the interface. This meant that the H&MBs and all other associated design deliverables were able to be issued on an accelerated schedule, again with high confidence in the product.

Inside KBR LNG's simulation, see Figure 3, a 'simplified' model of the Licensor process is normally used during the design. The details of this model are presented in Figure 4. This simplified model has to be adjusted every time that a change in the AGRU or the inlet conditions from the rest of the simulation is made.

Using the CAPE-OPEN interface of OASE connect, this process is much simpler, since the OASE AGUR unit operation is fully embedded into Aspen HYSYS®, see Figure . Therein, the OASE AGRU model (Figure 6) can be directly accessed, and any changes made in the simulation or in the AGRU are easily updated in the complete flow sheet.

Figure 7 illustrates the calculation sequence when a simulation is triggered in the PME. First all inlet information is transferred via the CAPE-OPEN interface to OASE connect, which triggers the calculation on BASF's server. When the simulation results are available in OASE connect, the results are automatically transferred via the CAPE-OPEN interface into the PME. This sequence will repeat in case of several iterations are running in the PME.

6. CONCLUSION

Digitalization of the work process of an engineering project allows a significant reduction of manual effort, decreasing engineering work hours and increasing quality due to avoidance of transcription errors. Prerequisite is that all engineering tools are interconnected. This paper illustrates how the engineering workflow can be facilitated by embedding a proprietary simulation tool into a PME via the CAPE-OPEN interface, shown here for the example of OASE connect. This embedding allows data to flow seamlessly between a CAPE-OPEN compliant PME and a CAPE-OPEN compliant proprietary simulation tool so that the user is able to produce complete and consistent H&MBs, allowing for the possibility of a digital process to create engineering documents and deliverables. In the inevitable case of changes in operating conditions or design parameters, the connectivity allows updated data to be automatically reflected in all connected downstream documents, further enhancing efficiency. The result is a savings in time and effort and facilitates effective work-sharing between offices worldwide.

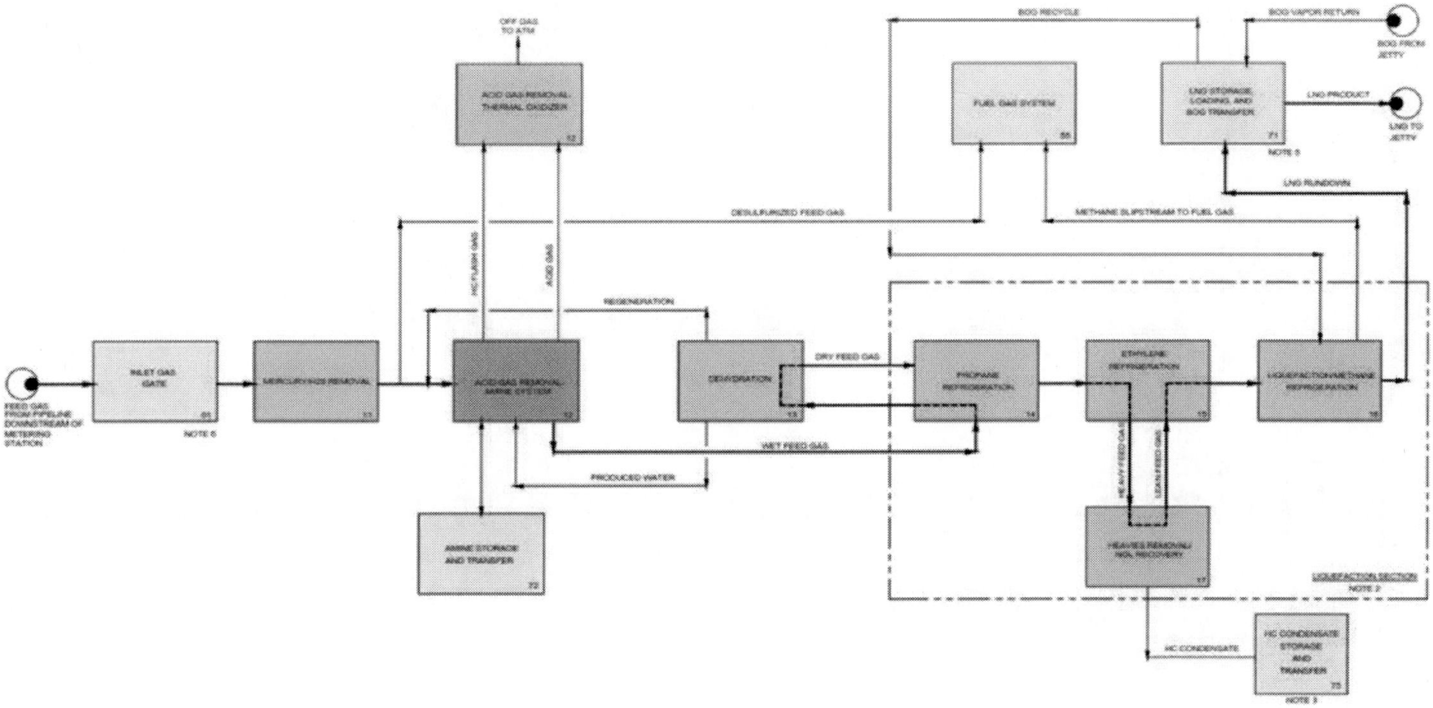

Figure 2: Block flow Diagram of example LNG plant

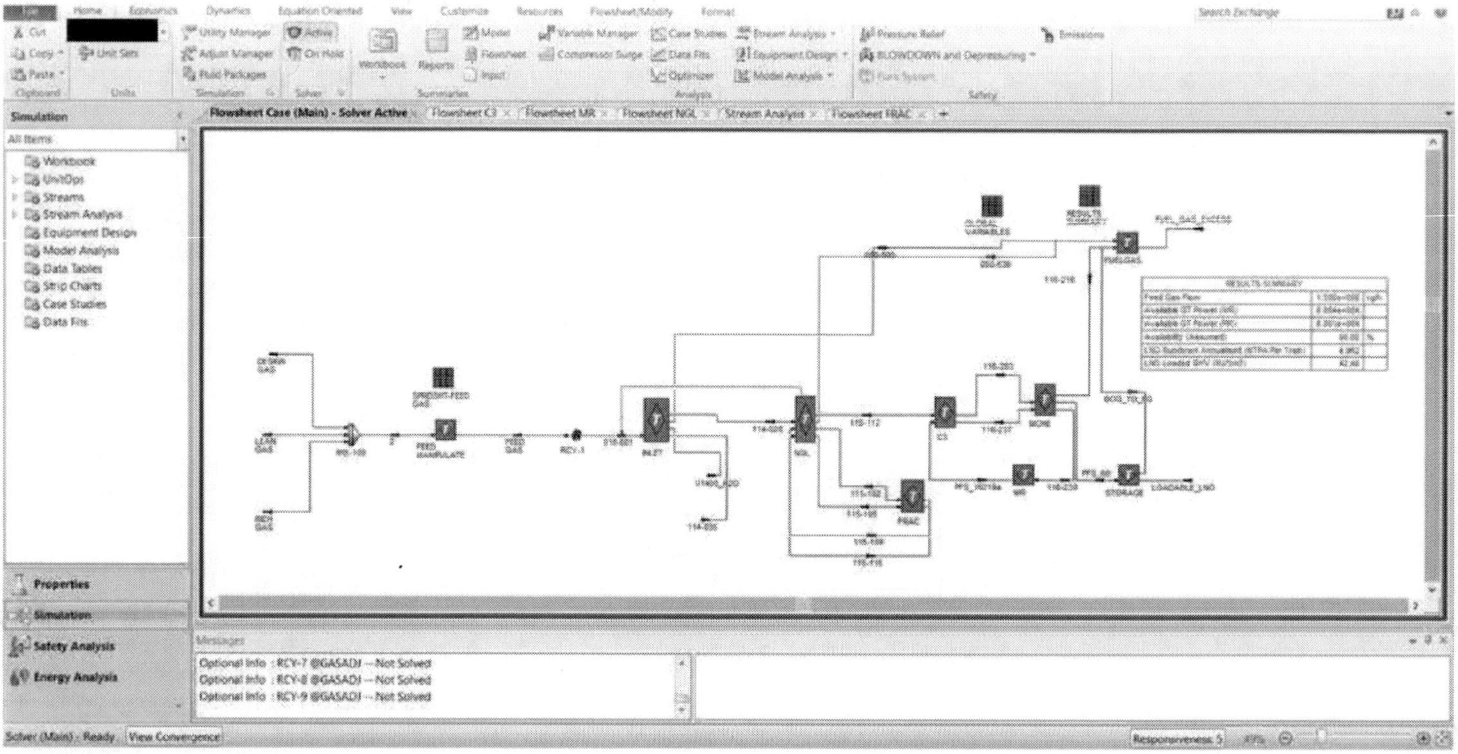

Figure 3: Aspen HYSYS® Model of an LNG Plant

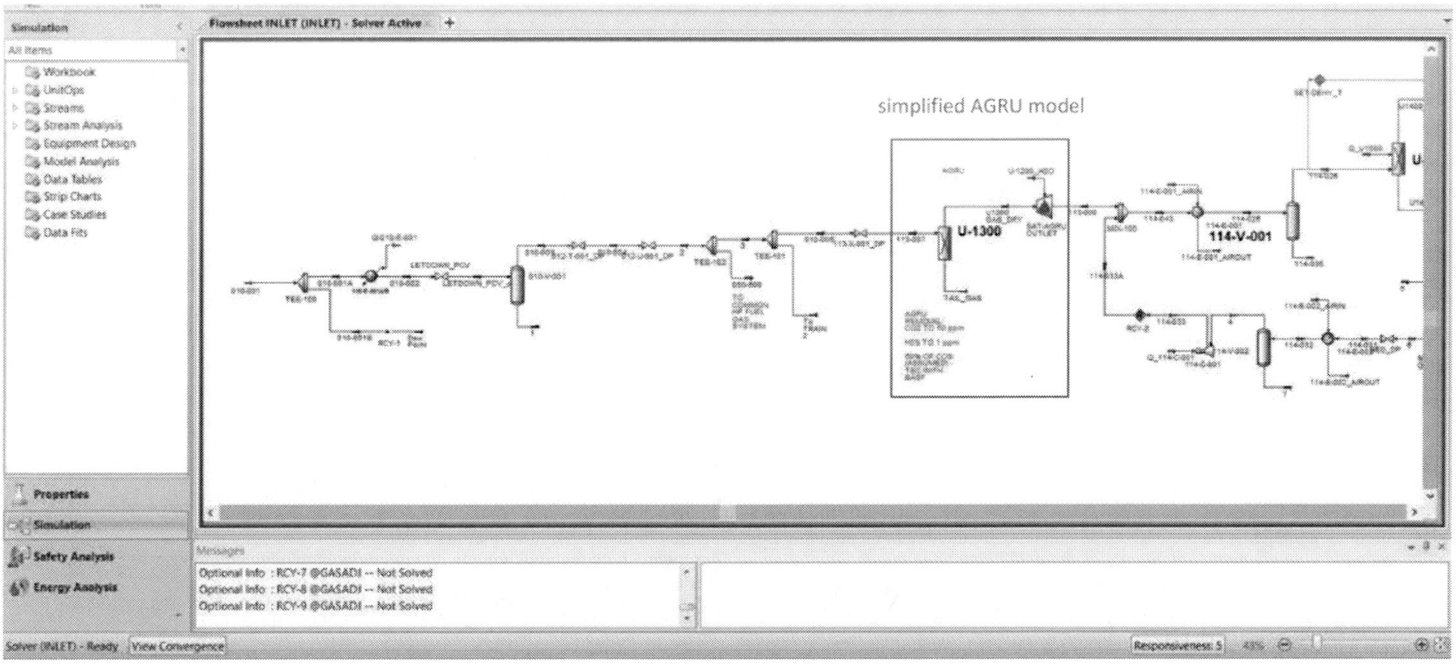

Figure 4: Inside INLET- AGRU Unit KBR Model

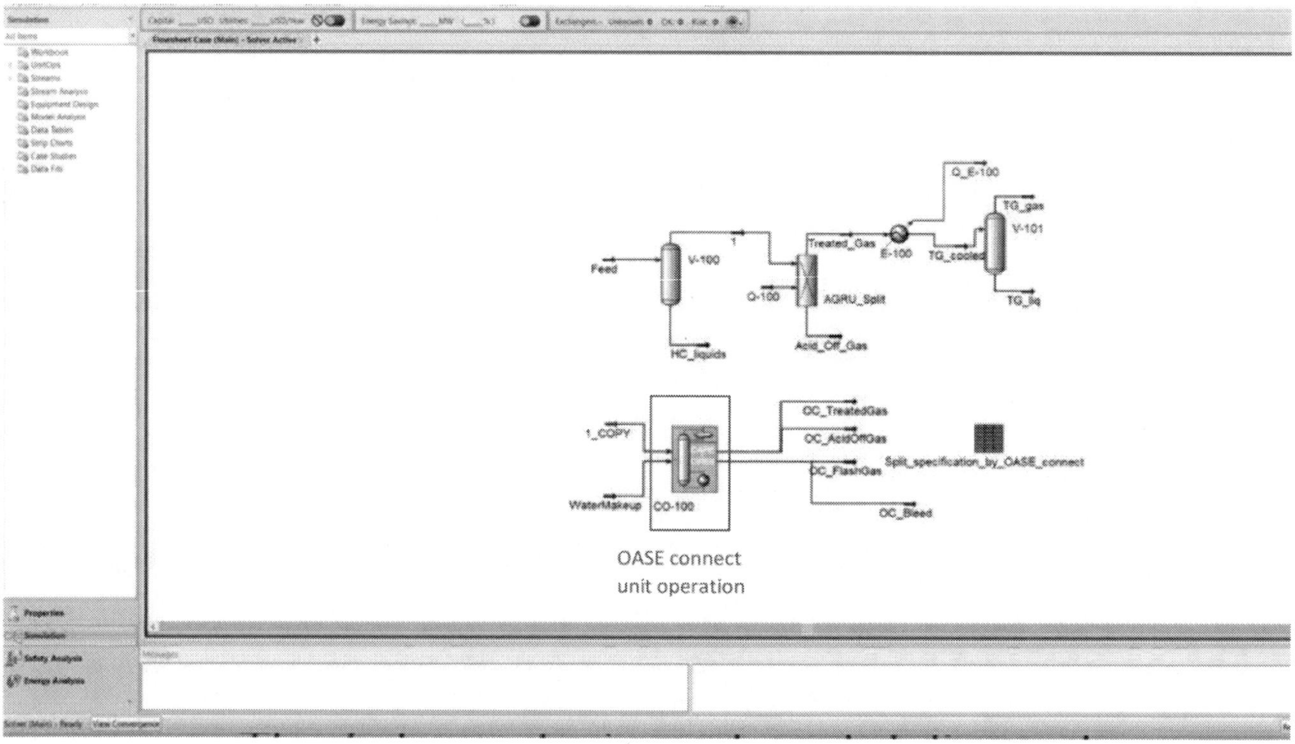

Figure 5: OASE connect unit operation embedded into a flow sheet in Aspen HYSYS®

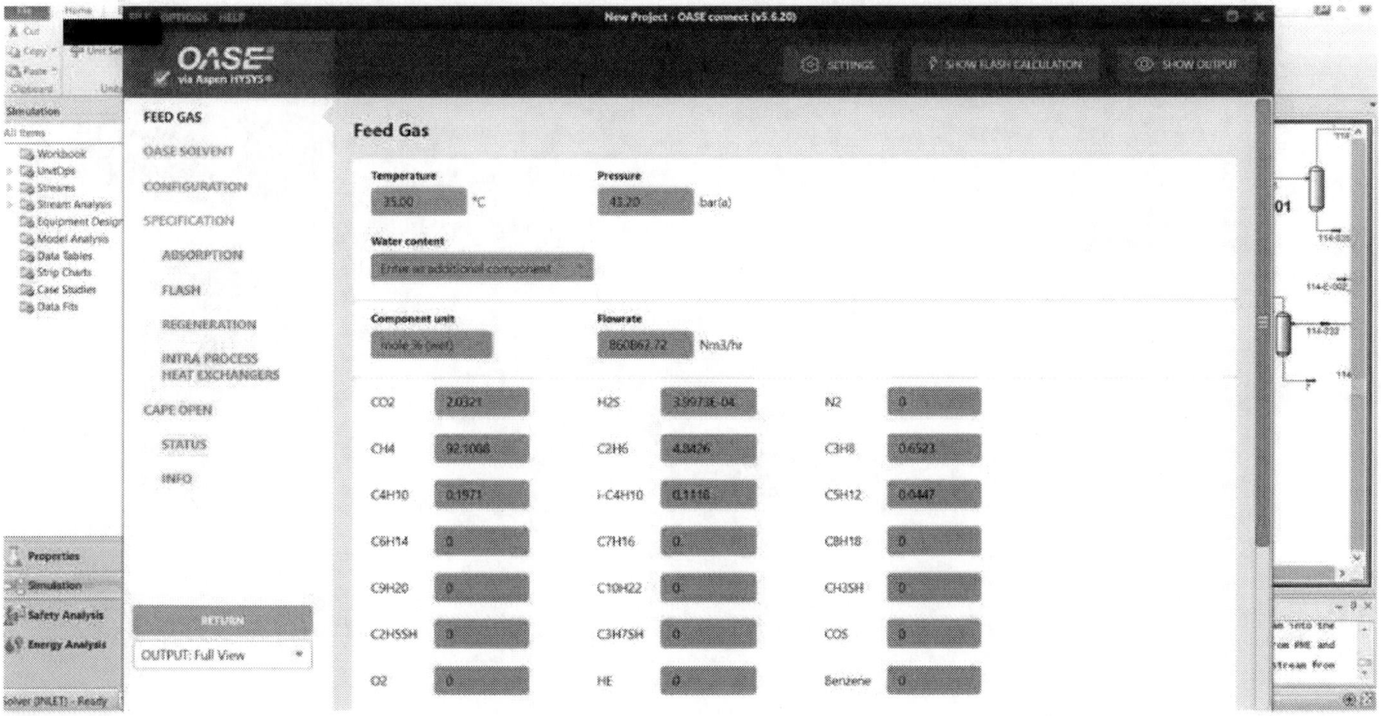

Figure 6: GUI of OASE connect inside of Aspen HYSYS®: Automatically transferred feed gas information

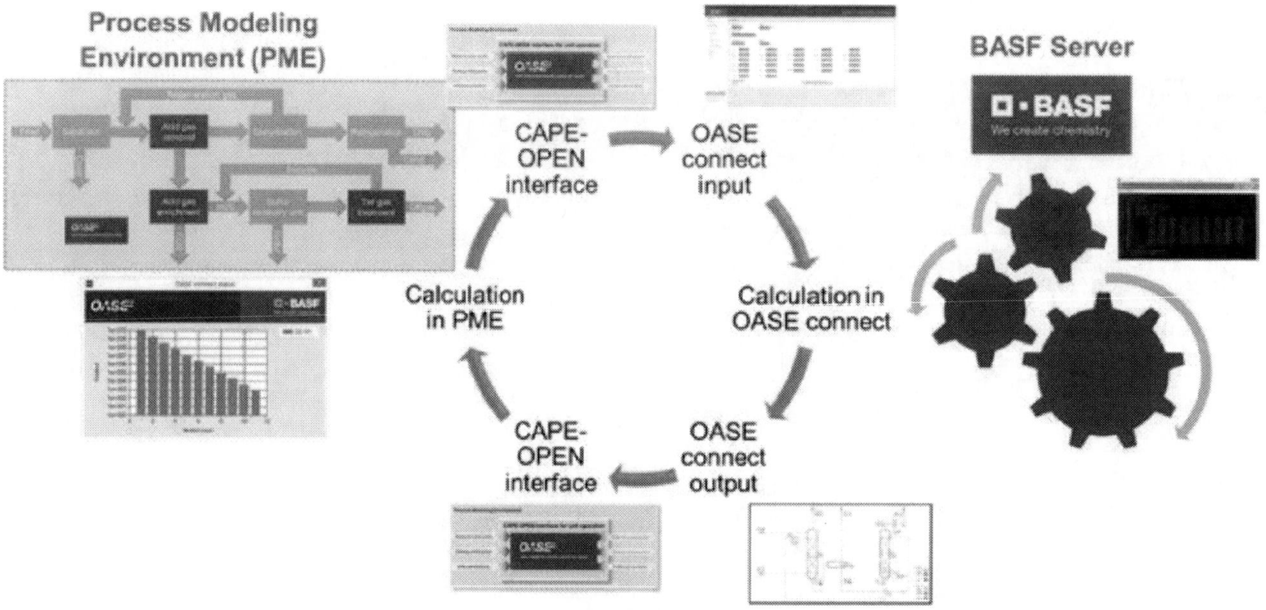

Figure 7: Simulation sequence between PME and OASE connect

590282 Multi-Scale Simulation: From Particle Resolved Fixed Bed Simulation to Multi-Tubular Reactor Design

Abstract:

Packed bed reactors are one of the working horses in the chemical and process industry. They are used amongst others for highly exothermic or endothermic catalytic reactions like (partial-)oxidation or (dry, steam) reformation of Hydrocarbons, which is typically done in multi-tubular reactors. These types or reactor are used since several decades but there is still room for improvement of the performance of these reactors and rigorous simulations can help here to achieve desired goals with lower upfront invest. Recent advances in modeling particle-resolved packed beds [1,2,3] allows a detailed inside in the flow, species and temperature distribution in the beds and therefore also into the conversion of the surface reactions. Based on these simulations, model parameters for simplified 1D-models can be estimated [4]. These simplified models can then be used to calculated multi-tubular reactors, either by running a computationally expensive simulation with resolved tubes or by coupling the CFD simulation to an advanced process simulation tool like gPROMS. In the later case, the flow non-uniformity as well as certain flow properties like coolant velocity and temperature is taken from the CFD simulations, while on the process modeling side heat transfer and reactions in the packed bed are calculated based on the simplified 1D models. This modeling approach is fully 2-way coupled and highly efficient in terms of accuracy and especially run-time and it can be embedded into a flow sheet simulation.

In this contribution we will show the whole simulation process and how the different steps intertwine with each other starting from the detailed particle resolved simulation all the way down to the flow sheet simulation. The benefit of this approach with be demonstrated based on several examples.

1. Eppinger, et al.; DEM-CFD simulations of fixed bed reactors with small tube to particle diameter ratios, https://doi.org/10.1016/j.cej.2010.10.053
2. D. Wehinger et al.: Detailed numerical simulations of catalytic fixed-bed reactors: Heterogeneous dry reforming of methane, https://doi.org/10.1016/j.ces.2014.09.007
3. Jurtz et al.: Advances in fixed-bed reactor modeling using particle-resolved computational fluid dynamics (CFD), https://doi.org/10.1515/revce-2017-0059
4. Jurtz et al.: Determination of Effective Transport Parameters from Particle-Resolved CFD Simulations for a Simplified Fixed-Bed Reactor Modeling (65a), AIChE Spring Meeting 2019, New Orleans

591047 Using Refinery Wide Simulation to Support Processing New Feedstocks from an Operability and Equipment Design Perspective

Abstract:

Introducing a new feedstock to the refinery is becoming more common in the search for greater profitability, however it can carry a high operational risk. The challenges do not stop with the feed reception tanks as the differences in chemical and physical properties can impact asset operation, equipment selections and maintenance priorities throughout the refinery.

In this session we will consider how refinery wide simulation can be used to evaluate the challenges of introducing a new (and very different) crude into the refinery. We will look at how changes in boiling range, contaminants and molecular species for the crude blend affect the operations of both the separation and conversion units. Process simulation, such as Petro-SIM, that has the range of kinetic refinery reactors models natively integrated is an essential tool to develop refinery wide models of enough process complexity. The session will look at benefits from digitalization of the technology and core developments in the accuracy and speed of solution.

The complete model can then be used to support operations, facilitate optimization and help with strategic (re)configuration studies. Once potential constraints and operational difficulties have been identified, we will show how simulation can be used to pinpoint strategies for relieving these constraints, whether by changing operating methodology or by capital investment. Where new equipment is required, we will highlight how Petro-SIM can be used for detailed design, showcasing how using a single simulation platform throughout the lifecycle of the project leads to improved consistency and more efficient work practices.

590284 Integrated Scheduling for Midstream and Downstream of Oil Refinery Supply Chain

Abstract:

Crude oil uploading, transfer, manufacturing, and oil-product distribution systems are critical operations and highly correlated in the entire oil refinery supply chain. However, very few efforts have focused on the integrated scheduling for the midstream and downstream of oil refinery supply chain. If they were scheduled separately, imbalances or discrepancies between the crude transfer, production and distribution sectors may happen, thus the loss of the profit margin across the refinery supply chain. On the contrary, the optimal coordination of these three systems could benefit not only the industrial profitability, but also the reduction of both management risks. Many published studies have addressed crude transfer, refinery production and pipeline distribution systems separately, while only a few works have addressed two or three sectors simultaneously. Existing publications either focused on the strategic and tactical decision level for the upstream and midstream of crude oil supply chain (Azadeh et al., 2017), or simply considered the scheduling problem with a discrete framework (Guyonnet et al., 2009, Tong et al., 2012). However, there is still lack of the systematic and integrated study dealing with the continuous-time and continuous-slug based extremely large-scale refinery supply chain problem.

In this work, a new continuous-time based integrated MINLP model is developed to optimize the scheduling of crude unloading and transfer, refinery manufacturing operation and multi-product pipeline distribution. The scheduling objective is to minimize the total operating cost; meanwhile, operating rules and product specifications, inventory limits, delivery constraints and oil product demands must be satisfied. The scope of this work contains crude unloading and transfer, refinery processing and multi-product pipeline distribution system. The crude unloading and transfer system includes vessels, storage tanks and charging tanks. The refinery includes crude distillation, reforming, cracking, hydrotreating, blending, gas processing, and sulfur recovering facilities. The long-distance pipeline distribution system connecting the refinery tank farm and multiple depots, which transport oil products to different local markets. Overall, the integrated simultaneously scheduling model could provide optimal solutions such as crude vessels unloading, crude transfer profiles, processing stream property and flowrate for unit operations, yield and inventory of various petrochemical products, injection sequence and volume of new batches, oil slugs movement inside pipeline, as well as oil product discharging to multiple depot tanks based on consumer demands. The efficacy of the developed integrated scheduling model is demonstrated by an industrial scale study.

Keywords: Simultaneous scheduling, Crude unloading, Crude transfer, Refinery manufacturing, Multi-product, Pipeline distribution, MINLP

References

Azadeh, A., Shafiee, F., Yazdanparast, R., Heydari J., and Keshvarparast, A. (2017). Optimum Integrated Design of Crude Oil Supply Chain by a Unique Mixed Integer Nonlinear Programming Model. Industrial & Engineering Chemistry Research, 56: 5734–5746.

Guyonnet, P., Grant, F.H., Bagajewicz, M.J. (2009). Integrated Model for Refinery Planning, Oil Procuring, and Product Distribution. Industrial & Engineering Chemistry Research, 48: 463–482

Tong, K., Feng, Y., Rong, G. (2012). Integrated Model for Refinery Production and Pipeline System Scheduling. Proceedings of the 22nd European Symposium on Computer Aided Process Engineering, 17 - 20 June 2012, London.

589956 Industrial Water Reuse

Abstract:

Water reuse in industry continues to expand, this presentation covers examples from oil and gas.

590325 Membrane Processes for Reuse Applications

Abstract:

There is an increased interest from manufacturers in various industries to recycle the wastewater (WW) they generate to reuse to meet their utility demand or even within their manufacturing processes. Typically, the water will need to be treated through many unit operations to purify it to a level it suitable for reuse. Membrane processes are nearly always required for one or more of these unit operations, but the design and operating philosophies of these systems need to take into consideration additional variables and challenges to ensure long term performance and reliability.

This joint presentation between a membrane technology manufacturer and system integrator will primarily focus on the design practices in the reuse processes for refinery WW and where ultrafiltration (UF) or membrane bioreactor (MBR), and reverse osmosis (RO) membranes fit in the overall process. Additionally, there will be some selected case studies highlighting a few examples of different industrial WW treatment projects that the presenters have been involved with. These will include examples in addition to refinery WW.

589771 The Next Frontier of Biological Nutrient Removal

Abstract:

The negative impact of nutrients such as nitrogen (N) and phosphorus (P) on receiving water bodies such as eutrophication and algal blooms led to development of biological nutrient removal (BNR) processes. Many of these initial BNR processes utilized existing infrastructure and involved alternate configurations and operational approaches for conventional activated sludge processes (ASP). Increasing population growth and subsequent impacts on climate and natural resources such as water have led to a need for more robust and sustainable approaches. As BNR technologies have continued to evolve to meet these ever-increasing demands, the next generation of solutions has focused on more efficient use of energy and oxygen and reduced footprint by introducing process intensification technologies. These technologies utilize attached growth and alternative biokinetic pathways to provide higher rate treatment in a smaller footprint or to provide different treatment pathways that are more favorable in terms of inputs (oxygen) or outputs (end products). These processes also provide advantages when treating recalcitrant or potentially inhibitory waste streams such as those found in petro-chemical applications. This presentation will encompass a brief overview of the conventional biological removal processes, a discussion of the optimization strategies, and a focus on looking forward to the emerging BNR technologies. Several of these next frontier technologies that will be discussed include membrane-aerated biofilm reactors (MABR), aerobic granular activated sludge, and short-cut nitrogen removal processes. As these technologies are transitioning into mainstream applications, the developers and the technical community continue to seek new applications in which their unique characteristics and advantages can be applied.

590252 Understanding Exchanger Practically for Improving Performance

Abstract:

Exchanger is one of the key asset in any operating unit and important aspect to ensure overall process performance improvement. Optimization of exchanger operation is critical and is feasible through application of right predictive and prescriptive analytics techniques.

A good monitoring process continuously tracks an equipment closely to identify the deviation in the operating parameters and understand issues before it start impacting performance. Nevertheless with multiple process parameter changes it's not easy, hence use of advanced analytical tools in combination with machine learning techniques and AI is valuable and effective to capture the deviations early and identify it correctly much better than traditional manual efforts. Pro-active technical approach always helps in anticipating the problem early before it can reach alarming levels and generally provides time for correction.

This paper will discuss techniques to identify deviation in equipment performance easily, accurately and get insights quickly. Additionally, the paper will describe how predictive and prescriptive analytics can help determine the impact of deviation, need for corrective action and assist in reaching challenging solutions.

Exchanger performance troubleshooting can be a tedious process as multiple operating and design factors affects the actual performance. At certain times it would require smart design improvements, but at times options are available to restore the desired performance as much as possible through intelligent process adjustments. Ensuring optimal exchanger operation will require complete understanding of the design, process complexities and overall functioning of the equipment.

This paper will cover some of Ingenero's case studies to demonstrate on how Manufacturers were able to gain substantially enhancing their Overall Asset Effectiveness by improving the exchanger monitoring and performance.

590609 Predictive Monitoring and Prescriptive Maintenance to Help Decision Making in Refinery Turnarounds

Abstract:

Predictive monitoring and prescriptive maintenance to help decision making in refinery turnarounds

Emilio Diaz-Bejarano[1], Manuel Martinez-Tafalla[2], Guillermo Galan[1] and Francesco Coletti[1,*]

[1]Hexxcell Ltd., Foundry Building, 77 Fulham Palace Road, London, W6 8AF, UK

[2]Repsol C.I. Cartagena, Valle de Escombreras, 30350 Cartagena, Spain

*Corresponding author: f.coletti@hexxcell.com

Refineries turnarounds (T/A) take place at the end of long operating cycles that may last for several years. This extended period of shutdown is used to carry out maintenance activities, revamps, inspection and testing of equipment or debottlenecking projects. Turnarounds are expensive due to the additional cost of labour, materials and contracting and loss of production. As a result, refinery personnel must carry out the planned activities efficiently and often quickly react to unexpected events to allow the plant to return into operation as soon as possible.

Predictive monitoring and prescriptive maintenance are powerful tools that can support decision-making not just during normal operations but also during T/A as they allow predicting the performance upon start-up and detecting inefficiencies of the process once the plant is in operation. In this paper, we present a case study showing the benefits of using such tools during the T/A of a Repsol refinery being monitored with Hexxcell Studio™.

During the T/A, the predictive capabilities of the monitoring tool were used to predict the temperatures along de train and the distance with respect to the furnace maximum capacity under different scenarios considered for the start-up by operations. This included: a) the possibility of starting up with some of the exchangers bypassed (as they were undergoing repairs); b) the impact that tube-plugging of one of the key exchangers in the network could have on the thermo-hydraulic performance of the network. The results guided the plant engineers to decide on configuration and timing of the start-up considering not only the performance of the train, but also possible effects in downstream processes.

After the start-up, the monitoring system allowed detecting inefficient performance of a heat exchanger that had passed unnoticed with standard monitoring methods. This allowed the engineers to react quickly, inspect the unit and fix the issue, leading to an immediate increase of the heat recovery in the network. The presentation will show the benefits obtained both in energy and economic terms as a result.

592114 Process and adsorbent for separating ethanol and associated oxygenates from a biofermentation system

Abstract:

Abstract for poster presentation or talk:

Abstract Title: Process and adsorbent for separating ethanol and associated oxygenates from a biofermentation system

Presenter Information: Deepak Sharma; Sr. Process Engineer, Engineering and Technology, Bayer AG

Patent number: US8658845B2

Patent Authors:

Deepak Sharma - Sr. Process Engineer, Bayer Ag, St. Louis; (Presenting Author)

Anil Oroskar – CTO, Orochem Technologies, Napervielle, IL;

David W. House – Scientist, Orochem Technologies, Napervielle, IL;

Alice M. Havill – Scientist, Lanzatech, New Zealand

The principles of process intensification if used on existing processes can lead to development and redesign of existing processes to become more sustainable. This presentation will discuss a process and an adsorbent for the separation of ethanol associated oxygenates from a dilute mixture of ethanol and associated oxygenates in water in the presence of organic compounds derived from a biofermentation process. After pretreatment, the separation is carried out in a simulated moving bed adsorption system employing a stationary phase adsorbent comprising fluorinated carbon or modified C18 silica gel selective for the adsorption of ethanol and associated oxygenates, such as 2,3-butanediol, with a mobile phase desorbent selected from the group consisting of methanol, ethanol, propanol, and methyl tertiary butyl ether. The process is useful for removing water from dilute aqueous mixtures of organic compounds comprising ethanol in dilute concentration in water and produced by fermentation, biomass extraction, biocatalytic, and enzymatic processes which are not economically recoverable by conventional distillation methods.

Keywords: Separation; Adsorption; SMB; Purification

Patent Assignee: Orochem Technologies LLC

592111 Pump safety, the undiscussed topic in chemical industry leading to several fatal incidents

Abstract:

Abstract Title: Pump safety, the undiscussed topic in chemical industry leading to several fatal incidents

Presenter Information: Deepak Sharma; Sr. Process Engineer, Engineering and Technology, Bayer AG

Pumps are one of the equipment's in the process industry that require a thorough review during HAZOP studies. The constant source of energy given to the pump, if not converted to kinetic energy to flow the fluid, can lead to heating of the pumped fluid. If the fluid keeps accumulating the energy provided by the motor, in a blocked-in scenario, it can lead to very high temperatures eventually leading to loss of containment of the pump and the pump can get damaged. There have been instances in the industry where fragments of damaged pumps have flown several feet away from the point of loss of containment leading to injuries. The poster will outline a scenario where a pump is run blocked in on suction and discharge for water and a chemical which has a low thermal decomposition temperature. Arc data is used to evaluate the onset temperature of thermal decomposition. The poster will outline safeguards that can be employed to avoid catastrophic loss of containment of the pump. Outlined below are safeguards used to avoid loss of containment of a pump in case of a blocked-in scenario:

- Install PSV on suction of pump
- Employ a low amp/power interlock on the pump to turn off the pump in case of blocked in situation
- Have a high-pressure interlock on the discharge line of the pump to direct flow back to the tank in case of pump deadhead
- Install an Interlock to turn off the pump in case the feed forward valves (automatic valves) close inadvertently

Pump deadhead scenarios are not only catastrophic when in flammable service there are several salts that have low thermal decomposition temperatures. Pumps in water service left blocked in could also blow catastrophically leading to reportable incidents. I feel this is an important topic to bring to the attention of students and professionals in the chemical industry. Poster will include pictures of actual installed pumps along with required safeguards in the field along with supporting P&ID depictions.

589440 New Kind of Linear Nanocomposite Applied in Improving Rheological Behavior of Crude Oil and Its Mechanism Studied through Molecular Simulation Method

Abstract:

In this work, a new kind of linear nanocomposite produced by carbon nanotube(CNT) as pour point depressant (C2-100) was developed by blending with MCNT nanoparticle and copolymer named Ethylene-vinyl acetate (EVA). Its effect of reducing yield stress and modifying the morphology of wax agglomerate were also evaluated. Besides, the synergetic effect of C2-100 nanocomposite with the magnetic field was also conducted. The mechanism of modifying rheological behavior of waxy crude oil was studied through molecular simulation method, which aims at understanding the interaction between this nanocomposite and wax molecule. The results indicated that C2-100 nanocomposite could reduce the yield stress of waxy crude oil more effectively than the copolymer (EVA) even at low concentration, as well as changing the morphology of wax crystals and agglomerate. Moreover, its effect of reducing yield stress becomes more obvious when the alternate magnetic field was applied. In addition, the results of molecular simulation also presented that the effect of improving rheological behavior of C2-100 nanocomposite is related to the adsorption behavior of wax molecules on the surface of MCNT nanoparticle, which is adsorbed with EVA molecules. In addition, the adsorption behavior of different fragments of EVA molecules can also influence the adsorption behavior of wax molecules on the surface of EVA-attached MCNT nanoparticle, which may influence the effect of C2-100 to improve the rheological behavior of waxy crude oil.

589683 Solar Energy Conversion and Noise Characterization Sensing in Building Integrated Solar Photovoltaic Devices with Ventilation

Abstract:

A study on solar energy conversion and noise characterization sensing in solar photovoltaic devices with ventilation is conducted. A building integrated parallel plate solar photovoltaic (PV) device was built with a pair of PV modules, a ventilated air cavity and an insulating back panel of plywood board filled with polystyrene installed in an outdoor test-room. Photovoltaic amplification was attained with power output from a potentiometer through rotation of its circular knob. A parallel plate PV device was studied for its
electrical parameters as Resistance-Capacitance (RC) electrical analog circuit. The effect of inductive and capacitive heating losses was considered in evaluating electrical characteristics of a PV device exposed to solar radiation. The characterization of noise interference, due to power difference of two intensities in a wave on PV Device is presented. Noise characterization of a wave is obtained depending on type of wave. The difference of two power intensities is due to transmission of light, sound, heat, electricity, fluid, and fire into a parallel plate PV device. Based on analysis of a solar PV device, a definition with measurement expression and unit for an electrical noise were coined. Standard definitions of noise sources, their measurement equations, their units and their origins under limiting reference conditions were also devised. A review for sensing noise of physical agents is conducted for sensing noise of physical agents (sun, light, sound, heat, fluid, electricity and fire). With integrated environmental control, these environmental parameters are to be characterized through various environmental monitoring sensors.

589945 Structural Characterization and Model Representation of Vitrinite-Rich India Coal By Using Advanced Analytical Techniques

Abstract:

The detailed investigation of the chemical structure of Karharbari coal (vitritine rich subbituminous coal from India) by utilizing advanced instrumental analytical techniques such as solid state ^{13}C nuclear magnetic resonance (^{13}C-NMR) spectroscopy, Fourier transform infrared (FT-IR) spectroscopy, X-ray diffraction (XRD), Flash-pyrolysis gas chromatography mass spectroscopy (Py-GC/MS), Laser desorption ionization time-of-flight mass spectroscopy (LD-TOF-MS), X-ray photoelectron spectroscopy (XPS), and high-resolution transmission electron microscopy (HRTEM). The Karharbari coal was an amorphous loosely organized lower rank coal with aromaticity 0.79. This was indicated by XRD results and also confirmed with ^{13}C-NMR results (aromaticity =0.84) and FTIR results (aromaticity = 0.82). The average molecular weight of the aromatic cluster is 507 amu by NMR result which is closer to the result obtained by HRTEM (MW = 530.46 amu). The presence of different heteroatoms like carbon, oxygen, nitrogen and sulphur with their functionalities, were determined by using a XPS technique. Different carbon/oxygen forms present in the Karharbari coal are (C-C) and (C-H) 68.53%, (C-O) 23.46%, (COO^-) 1.96%, (C=O) 6.05% and different nitrogen forms such as pyridine, pyrrolic, quaternary, oxidized and their distribution (mol %) are 19.36%, 45.68%, 31.25%, 3.71 respectively, and different sulphur forms were present i.e. corresponding to thiophenes, sulphones, sulphuric acid and sulphates with the molar content of 16.47%, 41.61%, 21.37%, 20.55% respectively. The construction of the individual molecules will be discussed in detail in the rest of this paper. The individual molecules were geometry-optimized and Materials Studio Amorphous Builder module was used to construct a three dimensional associated structure.

589479 Life Management of High Energy Piping with Highlights from ASME B31.1

Abstract:

High energy piping (HEP) systems – or as defined by the ASME B31.1 Power Piping Code, Covered Piping Systems (CPS) – are critical assets with respect to production facilities. A failure can have catastrophic consequences from a safety perspective and in terms of equipment loss and downtime. In fact, loss of high-pressure steam due to a failure in a main run of steam supply piping could shut an entire production unit down. Because of this and the requirements of the Code, HEP maintenance and inspection programs are established to ensure that the integrity of the systems is maintained in a safe and reliable condition throughout their lifecycle. There is no one-size-fits-all for the life management of these critical assets because there is a great variety in age, design, materials, and owner objectives. Life management of these assets is beneficial in that it provides a structured, systematic approach to creating and implementing a program including evaluating the current program (whether formal or informal), what the objectives of the program are (short and long term), what equipment is in place, and its history.

A HEP program put in place for overall life management is not synonymous with an inspection program, although inspections are an important part of an overall program. Inspections should be used in conjunction with analytical tools, real time monitoring, and laboratory examinations. The overall approach is life management – not inspection management – of the equipment.

This paper describes a HEP life management approach which includes identifying locations of interest, risk-ranking those locations, implementing a plan for determining the health of the highest risk locations, and taking actions based on the system health that has been determined. This paper also presents and discusses some of the requirements of the B31.1 code, which at the highest level states that each operating company shall develop operation and maintenance procedures for piping systems deemed necessary to ensure safe facility operations based on the provisions of the code, relevant industry experience, the operating company's experience (and knowledge of its facility), and the conditions under which the piping systems are operated.

590131 Optimization of C3MR LNG Liquefaction Process Using a Parallelized Simulator

Abstract:

Controlling the cost of energy-intensive processes such as C3MR (propane/mixed refrigerant) LNG liquefaction is a prime opportunity for optimization. Our work combines different C3MR models to develop an optimization that uses multiple independent variables, relaxed constraints, and flowsheet alterations to achieve an optimal operating cost. This work uses a rigorous model of the process in a process simulator rather than a reduced model of the process response. Use of a first-principles process simulator increases the accuracy of the model and therefore allows a more realistic optimization, at the cost of increased computing time. Parallelization (using multiple CPU cores) in the process simulator greatly increases the speed of the optimization calculations. CHEMCAD's optimization tools and multi-core technology make it possible to easily define the optimizations, run the optimizations efficiently, and generate global optima with a high degree of confidence.

590196 Modifying Fischer-Tropsch Reaction Sequence with Addition of Zeolite to Iron-Based Activated-Carbon Supported Catalyst

Abstract:

In addition to previously presented scientific work, modified reaction sequence of Fischer-Tropsch (FT) process was studied by addition of Zeolite ZSM-5 to an iron-based activated-carbon-supported catalyst. Main reactions involved due to FT catalyst are Fischer Tropsch Synthesis (FTS) and Water Gas Shift (WGS) reactions. Main reactions due to ZSM-5 are cracking, oligomerization, isomerization, cyclization, aromatization, conversion of alcohols to hydrocarbons, and coke formation. The waxy and linear structure FT products are converted to gasoline range, lighter hydrocarbons, iso-paraffins, and aromatics using ZSM-5. The catalyst ratio of 1:2, FT catalyst: ZSM-5, is used at 300 PSIG and 300°F. Separate-bed configuration, i.e. FT and ZSM-5 are separated in a reactor, arrangement is used for the reaction sequence analysis. Alcohols with carbon-chain from 3-5 produced by FTS are converted into hydrocarbons over ZSM-5. Waxy hydrocarbons with carbon-chain greater than 23 produced by FTS are cracked and isomerized to small n-paraffins and iso-paraffins using ZSM-5. The hydrocarbon with carbon chain from 7-10 produced by ZSM-5 cracking reactions are converted to aromatic products over ZSM-5. Produced gasoline to kerosene range n-paraffins are further cracked into smaller chain (1-5) iso- and n-paraffins. Some of the iso-paraffins are oligomerized and aromatized over ZSM-5 to form more aromatics. Further, multiple aromatic rings combine to undergo cyclization and dehydrogenation reactions over ZSM-5 to form coke. Water produced as a byproduct of FTS is combined with some of the carbon monoxide, which is the WGS, to form carbon dioxide and hydrogen.

590427 Lateral Retention Force Required for Sliding Saline Droplets over Silianized Silicon Surface

Abstract:

The lateral retention force required to slide a liquid drop on a surface is expected to increase with the time that the drop rests on the surface prior to sliding till it reaches a plateau. The molecular reorientation of the molecules at the periphery of the solid surface subjected to the unresolved vertical component of surface tension force causes this phenomenon. In this system, the surface tension of deionized water is increased by adding NaCl and is expected that the lateral retention force is increased proportionally with concentration. The retention force increased with increase in concentration reaching a maximum only to drop to a minimum with concentrations closer to saturation. That is, the lowest sliding force is required for both lower and higher concentrations of NaCl. This sudden decrease in observed lateral force is not a function of surface tension and the surface on which the drop slides. This phenomenon is associated with adhesion of salt to the surface and is supported by the drop sliding motion and depinning of the liquid from the solid.

590433 Study of Droplet Motion across Polymeric Substrates By Centrifugal Adhesion Balance (CAB)

Abstract:

On a given surface, the contour of the drop changes accordingly based on factors of the force applied, chemical structure of the fluid, drop size, and surface topography. With increasing force that is exerted on the drop, we determine the comprehensive trend for drop motion with respect to the controlled factors mentioned above. Drop motion is categorized into four phases and the point at where the initial displacement of receding line is observed is defined as the onset of motion. In this study, the onset of motion is affected by the factors mentioned above and aids in the determination for another value correlation known as Gs, known as Interfacial Modulus. The resistance to interaction towards the droplet as observed is quantified with respect to change in surface topography. This study therefore characterized drop motion and specifically onset motion based on controlled variables (factors). Work of adhesion is also quantified across different substrates to characterize the detachment of liquid from its substrate. All these parameters such as Gs, onset of motion, contact angles were analyzed with different polymeric substrates, drop sizes, and various fluids used.

590665 Population Balance Modeling and Spatially-Oriented Control for Crystal Size and Shape in Crystallization Processes

Abstract:

Crystallization is an important separation and purification process with relevant applications in the food, pharmaceutical, fine chemical and fertilizing industries. In these processes, modeling and process control for crystallization are focused on crystal size distribution (CSD) and shape of the crystals in order to make it possible to achieve the desired quality and purity of the product of interest. The control of crystal size and shape is of considerable importance for industrial crystallization processes, as crystal morphology has great influence on downstream steps and also from the point of view of product specification criteria. The consideration of crystal shape in modeling of crystallization processes is also important due to its impact on product effectiveness, such as the bioavailability and tablet stability in pharmaceuticals. Control of the crystallization process is can be facilitated with online measurements, which provide information to be used in the modeling of the process. However, if online measurements are not available, the strongly nonlinear dynamics of these processes makes difficult to establish relations between some inputs (e.g. temperature) and the final desirable characteristics of the crystal product. In this context, image analysis has been shown to be a suitable technique to monitoring the process [1].

In our present work, a population balance model-based scheme for controlling the size and shape of crystals is presented. The modeling was performed using a two-dimensional population balance (2D-PBM) in terms of the two characteristic dimensions of the crystals, describing the evolution of crystal size and shape distributions along the batch time. To identify the nucleation, growth and dissolution kinetics parameters using the 2D-PBM approach along the length and width directions, QICPIC-LIXELL's high speed dynamic image analysis equipment was used for the online measurement of the CSD, shape parameters and aspect ratio distribution of crystals during the experiments. In this way, the developed model allowed for the prediction of the crystal size and shape distributions during growth-dissolution cycles.

The developed framework was employed to obtain optimal temperature policies for controlling the crystal size and shape distributions that meet target criteria for size, morphology and productivity using the dynamic image analysis along the runs. Based on the concept of using measurements along batch crystallization processes as a trajectory in a phase space [2], a trajectory-endpoint control problem is presented with an oriented control scheme that is applied to the shape of the crystals. The scheme was used to control the batch cooling crystallization of potassium dihydrogen phosphate. The reported experimental data demonstrate the application of the control policies for temperature to produce crystals of the desired average shape and size in different batch times. Thus, using the first principle modeling coupled to the developed trajectory tracking control, the proposed framework has the potential to be applied to industrial crystallization processes for shape control and to be extended to polymorph control investigations.

References:

[1] Eisenschmidt, H.; Bajcinca, N.; Sundmacher, K. (2016) Optimal Control of Crystal Shapes in Batch Crystallization Experiments by Growth-Dissolution Cycles. *Crystal Growth & Design*, v. 16, p. 3297-2803.

[2] Griffin, D. J., Grover, M. A., Kawajiri, Y., Rousseau, R. W. (2015) Mass-count plots for crystal size control. *Chemical Engineering Science*, v. 137, p. 338-351.

590450 Auditing the Process Hazard Analysis - Key Characteristics to Scrutinize in a Good-Quality HAZOP Study and Supporting Effective HAZOP Study Updates

Abstract:

Although Process Safety Management (PSM), requiring good-quality Process Hazard Analyses, was promulgated in the United States in 1992, and HAZOP Study techniques date back to the mid-1970's, the HAZOP Study technique is still not uniformly applied in a high quality fashion. The objective of this paper is to provide experience-based tips on Process Hazard Analysis auditing, and some key characteristics that the auditor can scrutinize to assure quality implementation. Analogously, a similar set of characteristics that can facilitate future HAZOP Study updates can also contribute to a quality HAZOP Study that supports the spirit of Safety Management System objectives for the identification of hazards potentially impacting plant personnel, the community, and the environment. This paper will also review tips for structuring the HAZOP Study to facilitate later updates and revalidation.

590342 Development of High Performance Organic-Based Cathodes

Abstract:

For lithium-based batteries, structure/property relations in terms of cell voltage are desired for new organic-based cathode active materials. In this initial pursuit of structure/property relations, ab-initio calculations were performed on a few organic-based molecules to show the effect on the theoretical cell voltage by inclusion of nitrogen and/or oxygen. Preliminary test results are presented on a newly developed molecule and methods to potentially improve on the cell performance are given.

590705 Electrochemical Kinetics of Degradation of Graphite Anodes in Lithium Ion Batteries

Abstract:

Growth of the solid electrolyte interphase (SEI) is a major driver of capacity fade in LIBs. Despite its importance, the fundamental mechanisms remain unclear, primarily because of the complicated reaction pathways involved[1–3]. SEI growth can be both electrochemical and chemical in nature[4], and thus, it is a strong function of the potential and degree of lithiation of the electrode. We model the early-stage and long-term growth of SEI by accurately capturing the potential dependence of its formation kinetics as well as long term rate limiting steps, and validating it against the world's largest open source battery cycling data, generated in-house[5]. This is done using the Multiphase Porous Electrode Theory (MPET) framework[6] on graphite (phase separating) and carbon black (non phase separating) particles.

Lithium plating is another key degradation phenomenon that has been elusive, and it becomes important while trying to fast-charge batteries, i.e., 0% - 80% state-of-charge in 30 mins. We show that lithium plating is a key function of electrode morphology, phase-separation dynamics and potential. Phase-separation in graphite is modeled in the electrode using the Cahn-Hilliard Reaction framework described by Bazant[7]. We understand the electrochemistry of the onset of lithium plating with in-situ measurements connected to real time cell potential in a phase-separating electrode for the first time[8].

Results indicate that the peak SEI-forming currents are higher for higher driving currents. Also, we find that SEI only grows during electrode lithiation, i.e. the battery only degrades while being charged. We also find that onset of lithium plating is correctly captured only when phase separation in active material is accounted for. Further, the onset of plating is delayed on electrodes with a thick SEI layer – understanding SEI/plating coupling is integral to predicting fast charging manufacturing protocols for LIBs. This work holds promise for the predictive design of procedures[9] for manufacture and formation of LIBs.

[1] Cohen, Y. S.; Cohen, Y.; Aurbach, D. Micromorphological Studies of Lithium Electrodes in Alkyl Carbonate Solutions Using in Situ Atomic Force Microscopy. J. Phys. Chem. B 2000, 104, 12282–12291. https://doi.org/10.1021/jp002526b.

[2] Horstmann, B.; Single, F.; Latz, A. Review on Multi-Scale Models of Solid-Electrolyte Interphase Formation, Current Opinion in Electrochemistry 13, 62-69 2019.

[3] Nie, M.; Abraham, D. P.; Seo, D. M.; Chen, Y.; Bose, A.; Lucht, B. L. Role of Solution Structure in Solid Electrolyte Interphase Formation on Graphite with LiPF6 in Propylene Carbonate. J. Phys. Chem. C 2013, 117 (48), 25381–25389, https://doi.org/10.1021/jp409765w.

[4] Das, S.; Attia, P. M.; Chueh, W. C.; Bazant, M. Z. Electrochemical Kinetics of SEI Growth on Carbon Black: Part II. Modeling. J. Electrochem. Soc. 2019, 166 (4), E107–E118. https://doi.org/10.1149/2.0241904jes.

[5] Attia, P. M.; Das, S.; Harris, S. J.; Bazant, M. Z.; Chueh, W. C. Electrochemical Kinetics of SEI Growth on Carbon Black: Part I. Experiments. J. Electrochem. Soc. 2019, 166 (4), E97–E106. https://doi.org/10.1149/2.0231904jes.

[6] Smith, R. B.; Bazant, M. Z. Multiphase Porous Electrode Theory, J. Electrochem. Soc. 2017, 164 (11). https://doi.org/10.1149/2.0171711jes.

[7] Bazant, M. Z., Theory of chemical kinetics and charge transfer based on nonequilibrium thermodynamics, *Accounts of Chemical Research*, *46*(5), 1144–1160. https://doi.org/10.1021/ar300145c

[8] T. Gao, Y. Han, S. Das, T. Zhou, D. Fraggedakis, N. Nadkarni, C. N. Yeh, W. Chueh, J. Li, M.Z. Bazant, Interplay of lithium intercalation and plating on graphite using in-situ optical measurements, submitted.

[9] Huang, W.; Attia, P. M.; Wang, H.; Renfrew, S. E.; Jin, N.; Das, S.; Zhang, Z.; Boyle, D. T.; Li, Y.; Bazant, M. Z.; McCloskey, B. D.; Chueh, W. C. and Cui, Y.; Nano Letters 2019 19 (8), 5140-5148. DOI: 10.1021/acs.nanolett.9b01515

589787 Model Development of Solid Oxide Fuel Cell Thermal Performance Using Artificial Neural Network

Abstract:

NASA's Johnson Space Center has recently begun efforts to eventually integrate air-independent Solid Oxide Fuel Cell systems, with landers that can be propelled by LOX-CH4, for long duration missions. Using landers that utilize such propellants, provides the opportunity to use such systems as a power option, especially since they will be able to process methane into a reactant through fuel reformation. To ensure fuel reformation in the systems, Steam Methane Reformation (SMR) will be employed. Various lead-up activities, such as hardware testing and computational modelling, have been initiated to assist with this developmental effort including the development of dynamic empirical models using artificial neural network. Such a modeling approach has shown to be accurate (R^2>0.9) to predict the thermal and electrical performance for SOFC systems. Hence, such models could be used for system optimization and control design.

Impact of Spacer Nature on Rheological Properties of Cationic Gemini Surfactant

Abstract ID: 589650

Shams Kalam[1], Muhammad Shahzad Kamal[2], Shirish Patil[1] and S. M. Shakil Hussain[2]

(1) KFUPM, Dhahran, Saudi Arabia

(2) Center for Integrative Petroleum Research, King Fahd University of Petroleum & Minerals, Dhahran, Saudi Arabia

Surfactants play a vital role in Chemical Enhanced Oil Recovery (cEOR) techniques. Gemini surfactants are a unique style of surfactants having more water-soluble hydrophilic head and oil-soluble hydrophobic tail units. Both groups are connected by a spacer at or near to the head groups. Polymers are added into the surfactants to increase the sweep efficiency of the flooding processes. This paper presents the rheological properties of Surfactant Polymer (SP) solutions in which two recently synthesized cationic gemini surfactants with different spacer nature are used.

The rheological properties of novel SP solutions were determined using Discovery hybrid rheometer (DHR-3) from TA Instrument. The geometric configuration set for this work was concentric cylinder. All experiments were carried out at 80 °C, while the shear rate for steady shear viscosity measurements ranges from 0.001 to 1000 s^{-1}. 0.25 weight % of a polymer FO 4290 SSH was used with different concentrations of newly synthesized gemini surfactants for Rheological measurements. Two hours magnetic stirring and twenty hours retention time were provided to make uniform surfactant polymer solutions.

Both surfactants exhibited good thermal stability in seawater with no precipitation. Shear viscosity and storage modulus were measured as a function of shear rate and angular frequency respectively at different surfactant concentrations ranging from 0 to 0.5 mM, where 0 mM shows pure polymer. We discovered that counter ions also influence the rheological properties significantly. Research outcomes revealed that the novel SP solution with a benzene ring and chlorine counterion produces better performance in comparison with benzene ring and bromine counterion. Moreover, the effect enhances if we increase one more benzene ring in the spacer.

This study reports two recently synthesized SP solutions for cEOR processes in carbonate reservoirs. It presents the effect of counterion on rheological properties and improves the understanding of tuning the surfactant structure in making SP solutions to increase the flooding sweep efficiency.

590257 Real Time Energy Management Systems Implementation for Continually Optimized Performance

Abstract:

It is well known that energy management is key for the optimal operation of refining and petrochemical sites. Industrial energy systems are constituted by complex utilities networks that are interconnected and have complex interactions that in order to achieve it best performance need to comply with increasing environmental rules, push against operating constraints and adapt to altering power prices and rapidly changing equipment availability.

After the initial installation, management and continuous improvement of these systems present challenges not only from a technological but also from an organizational point of view.

The implementation of an advanced digital Energy Management System (EMS) which includes capabilities for multi-period optimized scheduling, real time optimization and monitoring, deployed either as cloud-based or on-premise solutions that could be acting on open or closed loop is known to provides large benefits.

Critical is the sustainability of the EMS in the ever-evolving energy supply and demand of refining and petrochemical sites. Commercial, asset and organizational factors all change and the EMS needs to adapt to the developing environment of each enterprise meaning the type of implementation, including appropriate management structure, culture, best practices and Key Performance Indicators (KPIs) needs to be continually improved.

The EMS implementation at the large and complex Eastman site in Longview, TX, will be presented as an example. It will provide insight on the initial successes achieved in real time energy management for both, cost optimization and energy related KPIs performance tracking. The case study evolves to discuss how technical asset improvements, once replicated in the EMS, provided alternative operating modes, large additional benefits and sustained optimal

Andrew Lammers, NN2

Eastman Longview

Pablo Montagna*, Cristian Rojas

KBC Advanced Technologies, Inc

589818 Analysis of Fuel Cell System through Project Based Learning for Application of Classical Thermal Fluid Concepts in Mechanical Engineering Laboratory Course

Abstract:

A hands-on project involving analysis of air noncircular channels in Fuel Cell to measure the pressure was introduced in a thermal fluid laboratory course. Although the students studied fluid mechanics as a theoretical course, this project gave them the opportunity to build an experimental setup from scratch to help showcase their hands-on skills and to apply their understanding of relevant concepts of fluid sciences. The results of the project included a successful test of a conceptual procedure, designed to measure and calculate pressure drop and flow rates across the anode and cathode sides of a fuel cell. The students enjoyed completing and learning from the project. They also provided very useful feedback on the project by incorporating a functional experimental system for the future laboratory project. This project based learning opportunity allowed students to build their confidence in what they are learning as well as provide a glimpse of what the real world may hold.

589438 Measured Versus Calculated Sour Gas Properties

Abstract:

Measured versus calculated sour gas properties

Abstract:

Recent Environmental agencies have restricted standards for sulfur emission from gas chemical processing. Sour gas can extremely impact the delivery and processing by corroding the equipment. Gas properties such as compressibility, density and viscosity are a major factor controlling prediction of future production pipelines, and production tubing and pipelines in gas industry.

Sour gas effect on all gas industry aspects usually gas properties are measured in laboratories but occasionally experimental values are not available due to complexity and time consuming hence a reliable method would be proposed.

We are assessing the accuracy of the industry standard methods in calculating sour gas properties, compressibility factor using the following correlations:

Drunchuck - Abukkasem, Hall – Yarbarough, Drunchuck- Purvis- Robinson, Londono et-al, Alanazi et-al and Azizi et-al.

Viscosity using empirical correlations such as Carr-Kobayashi Burrows, Lea-Gonzalez –Eakin and Dean- Stiel.

Density using equations of state Redlich Kwang and Soave -Redlick - Kwang.

This study shows that the best correlation for calculating compressibility factor and density of high concentration sour gas is Londono-et al, all other considered methods are not accurate. All other methods investigated include high error especially for high H2S concentration.

However the error in estimation of Z factor for most of the considered methods has great impact on estimated sour gas density. Therefore we recommend that sour gas properties be measured experimentally at high pressure and high temperature conditions.

587967 The Study of a S-CO2 Oxy-Fuel Combustion Power Plant for Low Grade Coals

Abstract:

Recently, it is very important to have reliable solutions that can maximize power generation efficiency of current power generation processes and decrease emissions of CO2 and other pollutants because there are increasing concerns about the global warming and the depletion of fossil fuels. Among the available carbon capture technologies and power generation technologies, a S-CO2 (supercritical carbon dioxide) cycle based oxy-fuel combustion process appears to be a promising mid-term solution for the production of electricity from fossil fuels. Since supercritical carbon dioxide (S-CO2) involves a rapid density change at small temperature and pressure ranges, the S-CO2 cycle can extract large amounts of energy from relatively small equipment and it is possible to minimize or simplify a power generation block and to obtain higher thermal efficiency than existing steam cycles used in power generation plants. Therefore, an indirect S-CO2 oxy-fuel combustion power plants for low grade coals was designed and analyzed by using ASPEN-PLUS. Also, the operating characteristics of the indirect S-CO2 Oxy-fuel combustion power plant were compared with those of conventional air combustion power plants.

590099 Characterization and Performance Assessment of Oil Entrapment By Commercially Available Surfactants

Abstract:

Oil spills occur frequently and are a threat to the health and safety of human society and the well-being of the environment and ecosystem. Remediating oil spills is an ongoing challenge. Surfactant based dispersing agents, such as the Corexit® product line, are widely used for remediation of oil spills mainly through direct application on top of the oil slick. This work encompasses the investigation of variables that influence the emulsification process and therefore the encapsulation of oil within the surfactant micelles. Of central importance are the conditions under which oil entrapment in emulsions even occurs. A central finding is that there appears to be a threshold shear rate for emulsifying oil with surfactants, below which no emulsification occurs. Different schemes for applying shear stress in this work included simply stirring, using a high energy colloid mill, and flowing the mixture through a high shear capillary. Also, the sequence in which the oil/surfactant/water (O/S/W) components are mixed impinges strongly on whether and how much oil entrapment occurs. To try to correlate shear stress and velocity profiles throughout the sample volume with entrapment efficiency and kinetic stability of these O/S/W systems, Computational fluid dynamics (CFD) simulations were conducted. Another important factor is the ionic strength of the aqueous medium, which greatly impacts kinetics of these emulsions, but with a smaller impact on the threshold at which any emulsion is obtained at all. Surfactants tested include Corexit® EC9500A, sodium dodecyl sulfate (SDS), Triton X-100, and cetyltrimethyl ammonium bromide (CTAB). The quantification of oil entrapment was performed through either thermal gravimetric analysis (TGA) or differential scanning calorimetry (DSC).

Designing a Native Model for Assessment of the Effectiveness of Advertising

Sorush Niknamian
PhD, military Medicine Department, Liberty University, USA
Email: so.niknamian@gmail.com

Abstract

Reviewing and evaluating the effectiveness of advertising can be used as a criterion to convert weaknesses into strengths and budgeting. Also, Identification of the factors influencing the effectiveness of advertising can lead to more professional design and implementation. In the banking industry, with the increasing number of banks and increased right of customers to choose, the role of advertising has become increasingly important to succeed in this field of competition; so it should be said that advertising can make the bank close to its goals when it is sufficiently effective. The current research aimed to identify and rank the effective factors on the design of a native model for evaluation of the effectiveness of advertising of the Eghtesad Novin bank in Iran. The statistical population of this research is all managers and experts of the Eghtesad Novin bank. According to the research literature, the criteria and sub-criteria affecting the effectiveness of advertising have been extracted. By using the fuzzy Delphi technique and based on the experts' opinion, the final indices were extracted. Then, using the fuzzy AHP method, the rank of criteria has been addressed. The derived criteria are demographic characteristics, individual emotions, and attitudes of the customer, apparent features of the message, message content, market characteristics, cultural, economic features, design and nature of the service, advertising methods. The results of the ranking showed that the methods of advertisement, the design, and nature of the service, the apparent characteristics of the message, the emotions and attitudes of the customer, the demographic characteristics, the content of the message, economic, market characteristics, were placed in the first to ninth ranks, respectively.

Keywords: banking services, banking services marketing, advertising, advertising effectiveness,

Introduction:

Advertising is one of the promotional mix elements of marketing mixes, which play an important role in attracting customers. Advertising is a bridge between the organization and customers and can introduce the organization and its products to customers, inform consumers about the goods and services of the organization, convince them about the benefits of the product, and ensure them to have the best choice even after the purchase when the customer is uncertain (Buck, 2016: 172).

Given the increasing competition among the brands in the marketing sector, advertising is a necessary tool to introduce the characteristics of goods and services to customers. Leong et al. (1999) pointed out the fact that consumers overcome advertising information, and many of them do not remember the advertisements that they are exposed to, and it is necessary to consider the advertising and its attractive methods to attract customers' attention. The ability to attract customers' attention is one of the main indicators of advertising effectiveness that should be taken into consideration. The attention of customers is a key component of the overall effectiveness of advertising (Pileliene et al., 2017: 2). Any professional marketing without advertising will not be effective. Despite relatively high advertising costs, advertising is mentioned as an investment rather than the cost. Since many advertisements cause that the advertising companies do not benefit their advertising because of the wrong way of advertising, the competitive companies benefit from the advertising of that company (Hartnet, 2016: 23)

Advertising should be in line with the goals and plans, as well as the strategies of the organization. Therefore, we should see to what extent the advertising has met such goals. Allocating funds without evaluating the effectiveness of past advertising is the waste of resources of the organization. By evaluating the effectiveness of advertising, changes can be made to the form, the content of messages, type of media, budget, communication channels, and even the time and conditions for conducting advertising to create more useful and effective advertising than the past (Yazgirudo-Yasteh, 2015: 359). In the current competitive world, the importance of advertising is increasing. As a result, businesses have devoted a great deal of money to customers' attraction, but it is important to see whether the company advertising has been designed and implemented in such a way to meet the goals of the company. This issue has been often neglected in our country. (Kim, 2014: 2107).

The amazing effect of advertising on the lives of individuals requires that this scientific, social and communicative branch be carefully studied and advertising is made with a scientific, artistic, specialized and expert view. Professional advertisement can be a sharp sword for sales promotion, both inside and outside the country, and otherwise, it is a waste of time, money and energy. A professional advertising lever can, with the help of new technology, penetrate humans, and cause more desire for consumption (Arribeh, 2016: 463). Today, banks have turned their attention to the widespread use of mass media advertising, given the highly competitive structure governing on them. Also, given the relationship among a wide range of businesses of organizations and households and banks through various deposit accounts, loans, and other financial services, banks can benefit from advertising in a proper way to meet their goals, namely the introduction of their services, the communication with customers, their effectiveness on them in choosing a bank to do their banking business and ultimately attracting them. Unfortunately, in most banks, despite the massive advertising and the allocation of large amounts of energy and budget, the effectiveness of such advertisements is less evaluated. Since advertising is one-way, it is necessary to receive feedback from the customers (Rabiei, 2011: 19).

Regarding marketing, the effectiveness of advertising depends on the wide range of coverage of various consumers and the cost of advertising campaigns. Economists have defined the effectiveness of advertising as a change in the profit derived from promotional activities. Therefore, the effectiveness of bank advertising is the potential quality of information reported to consumers, and in promotional campaigns, it aims at creating a bank image and informing banking products and services, as well as obtaining a specific economic result before and after the transmission of the message (Denis et al., 2012: 85).

Eghtesad Novin Bank (EN bank), which is one of the oldest banks in Iran, due to the addition of new banks, should use new tools to protect and even increase its market share for the assessment of the effectiveness of its advertising. For this purpose and given that the bank uses various advertisements, it is necessary to design a model to evaluate the effectiveness of the bank advertising; therefore, the present study seeks to answer the following questions:

- What indicators are used to assess the bank native advertising model?

- How are the indicators for bank advertising evaluation weighed and ranked?

In this research, we have tried to identify the criteria using the Fuzzy Delphi process; then we will use the Fuzzy AHP method to rank the identified criteria. In the next section of the research, the literature has been studied. Then the research design has been expressed, and in the last section the conclusion has been presented.

1-Research literature:

1-1-Marketing of banking services

The growing trend in service activities has led to the emergence of new markets and large service companies. The complexity of service industries, such as information systems, management consultancy, and logistics, has increased, as service providers have entered the market with a variety of business backgrounds and capabilities to provide a broader range of services. Facing this complexity requires knowledge and skills to understand the services. The definition of the service includes all activities that have been done (for example, writing document specifications) (Selviaridis et al., 2013: 1398).

In the industrial literature, there are various ways to define the service and product requirements. Van Weele (2010) has distinguished technical and applied definitions from each other. The first requirement expresses the exact technical specifications of the goods with the required services and inputs while it focuses on performance demands and demand-oriented performance orientations. In this regard, Axelsson and Wynstra (2002) have proposed four methods for definition of services: input definitions are based on the sources and capabilities of the service provider while the process-oriented approach focuses on how services are presented regarding activities and processes. Output-oriented definitions focus on service capabilities and performance levels, as well as the interaction between the suppliers and recipients. Ultimately, they focus on stress-oriented results in the economic value of services (Selviaridis et al., 2013: 1399).

The financial sector forms an essential part of the service sector. In this regard, banks have used marketing to improve the strategic performance of banks. In particular, since the 1980s, this trend has been confirmed and criticized. Marketing researchers and experts are interested in researching the various aspects of marketing in banking. The quality of services is essential for customers who use banking services, and the provision of accurate and high-quality services leads to the survival of banks in the competitive market. Improving the quality of services will increase the access and response to the demands of the bank's customers, which will make the bank more distinct than other competitors. Therefore, measuring the quality of services in the banking industry is essential. Providing quality services is a hard and difficult task, in this regard, the features of services in general and banking services, in particular, should be explained. Banking services require a high degree of appropriate communication between service providers and customers. If services are not provided properly, they will be at high risk.

About the five features that distinguish them from material services (intangibility, non-allocation of materials for production, diversification, non-corruption, and lack of ownership), financial services have two other specific features, that have been firstly developed by McKechnie (1997). These characteristics include trusteeship and two-way information. These special features mean that the evaluation of services is more difficult than material products, and they not only create a challenge in the development of marketing activities, but also they are the source of the problem in providing an optimal level of quality services. Banking services require a high level of communication between service providers and customers, as well as a high degree of customer participation in the process of providing services. Angurr et al. (1999) emphasized that due to the high standardization and the development of technologies in the banking sector, human factor has been considered as the most important one in making distinctions among competitors. The expertise and motivation of the employees, as well as the level of customer involvement in the provision of services, significantly affect the quality of services. Due to the wide range of banking activities, attention should be paid to cultural and social changes that affect the attitudes of customers (Durdana et al., 2015: 100).

Marketing of banking services has three components as follows: if any of them were properly executed, banking services would change the country.

1- Market study: Identification of the banking services market system, including the status of current banking services and their weaknesses, strengths, expectations, demands, ideas and opinions of customers, the conditions of environmental instruments affecting the market system, in particular government policies and governing laws, and even global changes in the banking system and successful models in this system. Indeed, market study is the foundation of the evolution of the banking system, because market study (which is the result of constant communication with the market and conducting marketing studies and research in various forms) leads to comprehensive, up-to-date, relevant and necessary information for decision-makers and bank managers (Munsung Rheea & Satish Mehrab, 2006: 508).

2. Market creation: It involves the use of a variety of simple and complex marketing techniques and competitive strategies, as well as the growth, development and market creation in the banking system; it needs to understand the internal customers (employees) and external

customers, to communicate with them and meet their needs and demands. Market creation in the competitive environment requires appropriate strategies (Moch, 2013: 2908).

3. Market protection: One of the most important tasks of marketing in the banking system of the country is to maintain market share, a stable relationship with customers, and to convert ordinary customers to loyal and supportive ones. The more intense the competition, the more valuable the loyal customers will be, as marketing research in the world has shown that the cost of finding a new customer in firms is 3 to 33 times higher than the cost of maintaining current customers regarding the type of products and services. Therefore, market protection prevents the costly advertising, and attraction of new customers and provides a kind of confidence and serenity for market survival, especially in competitive conditions (Fredriksson & Moro, 2014: 69).

1-2-Effectiveness of bank advertising

One of the indicators of success in any organization is the extent to which the organization achieves its goals. Here, the success of any organization is determined by its own goals. This approach has been called effectiveness in the management literature. The effectiveness is the degree to which the organization achieves its goals. Goals must first be defined to measure effectiveness in any organization. Then, plans are set up to fulfill these goals, and thus the plans will be implemented. In the end, the results are compared with predetermined goals (Manjegani et al., 2015: 124).

About 100 years ago, William Heketh statement, "Half of the advertisements were wasted, we just do not know which half it is" attracted everyone to the accountability of advertising. What matters to advertisers is that how much the amount of money spent on advertising has returned? Has the cost spent on making and broadcasting the advertising been effective? Has the effect of advertising been strong or weak? Knowing the answer to these questions is important to the advertiser because today companies pay much money to make and broadcast advertising. According to Jack Trout, the money we use to advertise is very much. Every second of the World War II was $ 9,000, and Vietnam cost $ 22,000 per minute. But in the American Football Final, every minute of advertising cost $ 2 million, so the advertiser has the right to know about the effectiveness of his advertising (Aksoy, Atilla, 2005, Yeni Reklamcilik, Istanbul Bilgi Universitesi, 95). It is very difficult to measure the impact of long-term advertising on many occasions. Also, academically in the world of advertising, it is impossible to create a comprehensive theory that can be used to determine the effectiveness of advertisements on people at different levels under different circumstances. The effectiveness of media is its capacity to create desired results, or special conditions in a particular position and content, that is, attaining the desired communication goals (Wery, 2012)

Evaluating the effectiveness of advertising indicates how much advertising has succeeded in achieving the goals. The evaluation of the effectiveness of advertising acts as a feedback, informs us about the weaknesses and strengths of our advertising by which we find our shortcomings in the advertising. The marketer may choose an appropriate stage regarding product life or brand position, and use it to define his advertising goals and ultimately evaluate its effectiveness (Eckler & Bolis, 2011: 2).

The effectiveness of advertising is a function of its content (message), the execution (how it is transmitted), and repetition (the number of times each consumer sees the message) (Cutler, 1997, quoted by Moorthy & Hawkins, 2005). Some of the effectiveness criteria are the call and recognition of the brand name (Park, 2006), awareness, willingness and intention to purchase (Decrop, 2007), attitude towards the ad, attitude toward the brand name and intent to purchase (Muda et al., 2014)

The control of the communication element has a direct effect on customer perception of different dimensions of the brand name, such as service quality and loyalty. It means that the elements of communication, advertising and promotion must be carefully implemented and controlled (Suvittawat, 2015). Informative advertising is one of the factors that shape the behavior of the bank customers. Studies conducted both on attracting, retaining and loyalty of customers, show the need to examine the impact of advertising on attracting and loyalty of customers (Ioannou, Boukas and Skoufari, 2014: 25).

An ideal model of world-class advertising can: 1. transfer a consistent meaning to the consumer all over the world. 2. Transfer positive global capacity for the brand. That is, the results of advertising among different peoples should produce a positive effect and transfer high-quality signals (Strebinger et al., 2018: 410).

Advertisers want to prepare consumers to like goods or services, but sometimes it harms them through unwanted and intrusive advertising. It leads to the creation of a series of negative emotions in advertising. The purpose of ads is to create positive feelings in consumers about advertised goods and services (Bell & Buchner, 2018: 1).

On the other hand, research shows that commercial advertising can be positively evaluated and considered, and this positive evaluation leads to a positive evaluation of the brand with the product. However, advertising can have negative effects that lead to the formation of negative attitudes towards that product and service. Therefore, by increasing the volume of messages, it can have positive, negative or neutral effects. Therefore, marketers should evaluate advertising messages and their behavioral outcomes. With this regard, the effectiveness of advertising, based on a hierarchical approach, should be fully investigated (Tafani et al., 2018: 125).

There are various factors which influence effectiveness of advertisement like media, time, market life stage, advertisement life stage (new or old advertisement) and also advertisement creative signs (like content and administrative elements) (Talis et.al, 2005). Also using popular cartoon personages (Ranjbarian, 2009), size (Rousat, Venous & Ebrahimi, 2004, Baltas, 2003), color (Park, 2008, Moriarty, Mitchell, Wells, 2009, Cheng & Kao, 2011), complexity or simplicity of the advertisement (Morrison & Dainoff, 1972, Di Res, 2008), using irony factor (Mukherjee, Hober, 2004), using famous artists (Rousta, Venous, Ebrahimi , 2004) and consumer features (Matrin, Phaymi, Agee, 2002, Park, 2008) are among these variables. Advertisement repetition is also a factor which helps in reminding its content to the audiences (Rousta, Venous, Ebrahimi, 2004). One of considered factors in the debate of evaluating effectiveness of advertisement is the attitude towards advertisement. Attitude towards the advertisement includes public interest or lack of interest in an advertising stimulus in encountering a special topic (Moaen & Minor, 2009). Mental imagery (B one & Elen, 1992), content transparence and emotions called through the advertisement are among variables influencing the attitude towards the advertisement. Some of these factors may lead to strong impacts while some of them may create weak impacts or don't include any impact. Awareness of the impact of each of these factors makes a considerable help in making management decisions (Talis et.al, 2005).

Fotini et.al (2009) provided a model of effective advertisements for advertising websites of big companies. The results of the study showed that measurements, activities, priorities and intentions of people have a considerable impact on evaluation and qualitative factors of relation-oriented marketing have a great impact on advertisements effectiveness indicators. The extracted criteria include:

- Advertisement policies: product information, company information, product general display, consistency with traditional advertisement

- Medium features: message features, information structure, responding time, facilitating research, beauty, interactions, communications in real-time, amusement, information control

- Relation- oriented marketing policy: communications, feedback, supporting customers

Today, organizations for competition in the business world and selling their goods and services require purposeful and efficient marketing activities. One of topics which has a very important role in the process of marketing of an organization and assign a high volume of budget to it is advertisement. Advertisement causes some changes in the knowledge, attitude and behavior level of audiences, so attention to evaluating advertisement effectiveness is a necessity for the organizations that consider advertisement as requirement of continuous life, growth and development. By assessing advertisement effectiveness and investigating consistency with the organization objectives and its expected impacts, some changes may be created in its dimensions, content and methods so that a more effective and useful advertising is performed. Regarding the research literature and internal and external background which has been stated, here a general categorization of the model proposed criteria has been provided in the following table:

Table 1- summation and extracted variables

scholar	sub standard	criterion
) Dargi & Mohammadian (2010), Ashouri (2016), Mansouri Moayed (2005), Aghapour (2016), Mirzadh & Mehrmanesh (2016), Keramforn (2011), Mac Ki- Nesbiat et.al (2011), Rosemary et.al (2012), Wang (2015), Maryion et.al(2017)	awareness of sexuality differences	individual factors
Dargi & Mohammadian (2010), Ashouri (2016), Samsam Shariat Keramforn (2011)(2007), Mirzadeh & Mehrmanesh (2016), Mac Ki- Nesbiat et.al (2011), Rosemary et.al (2012), Wang (2015), Maryion et.al(2017)	attention to age requirements	
Dargi & Mohammadian (2010), Dehdashti Shahrokh (2009)	various consumers (permanent, ordinary and case consumers)	

Khashizadeh (2010), Danaei & Dasar (2016), Pillay et.al (2012), Marion et.al (2017)	advertisement richness	
Danaei & Dasar (2016), Dargi & Mohammadian (2010), Pillay et.al (2012), Olson et.al (2012)	advertisement accuracy	
Ashouri (2016), Rabiei (2011), Taeini Harari (2014)	positive mentality of the organization	
Dargi & Mohammadian (2010), Shahmohammadi (2011), Olson et.al (2012)	honesty feeling in advertisement	
Fakhari et.al (2012)	threatening feeling	
Ashourizadeh (2016), Hamidizadeh (2013)	attitude towards brand	
madhoushi (2013), Olson et.al (2012)	being convinced	
Ahmadi (2013), Mac Ki – Nesbiat et.al (2011), Olson et.al (2012)	convicting audiences	
Ashouri (2016), Hosivar (2014), Olson et.al (2012)	mental imagery	
Siadat & Sakhi Jourashk (2014)	positive feeling	
Navaei et.al (2015), Rozmari et.al (2012)	audience being active	
Mosavi Esmaeilzadeh (2015), Olson et.al (2012)	knowledge level	
Dargi & Mohammadian (2010), Ashouri (2016), Shafiei Nikabadi & Zarei (2017)	attitude towards advertisement	
Dargi & Mohammadian (2010), Hosseinzadeh et.al (2016), Sanei et.al (2016)	familiarity with methods and forms of advertisement	message
Ashouri (2016), Olson et.al (2012), Alan (2012), Patric Ro (2014)	attention to reduction of expectation time of the user for observing advertisement	
Danaei & Dasar (2016), Forentebury & Mac Gldric (2011)	beautiful and appropriate	

	combination of colors	
Aghapuor (2016), Sanei et.al (2016)	site and location of advertisement placement	
Hassanzadeh & Danaei (2016), Sanei et.al (2016)	access to sufficient and required information	
Dargi & Mohammadian (2010), Darzban (2016)	starting the advertisement with a challenge	
Navaei et.al (2015), Salimi Fard & Pourdehghan (2016), Bargar et.al (2016)	starting the advertisement with a statement and explanation	
Dargi & Mohammadian (2010), Mirzadeh & Mehrmanesh	using helping words and terms for decision making	
Dargi & Mohammadian (2010), Daneshvar et.al (2016), Alan (2012)	using stimuli with diversions more than usual and expected extent	
Sanei et.al (2016)	using images of women and children	
Sohrabi Ranani (2016), Fortenbury & Mac Goldric (2011)	creativity in the message	
Shafei Nikabadi & Zarei (2017), Denis et.al (2012)	the text of advertisement message being a predicate sentence	

Khashizadeh (2010), Dargi & Mohammadian (2010), Sohrabi Ranani (2016), Patric Ro (2014), Abdolsalam et.al (2016)	continuation and repetition of message	
Daneshfar & Nabizadeh (2015), Fortenbury & Mac Goldric (2011)	using trustable and famous people	
Salimifard & Pourdehghan (2016)	common dialects and terms	
Daneshfar & Nabizadeh(2015), Forentnburi & Mac Goldric (2011)	loveliness of the popular person for the observer	
Daneshfar & Nabizadeh (2015), Forentnburi & Mac Goldric)(2011	familiarity of famous person with the observer	
Mansouri Moayed (2005), Aghapour (2016)	emotional music	
Khonsazadeh (2010), Hosseinzadeh et.al (2016), Aghapour (2016)	using various methods	
Siadat & Sakhi Jourashk (2014), Sohrabi Ranani (2016)	high coverage and being universal	
Navaei et.al (2016), Mosavi Esmaeilzadeh (2015), Hosseinzadeh et.al (2016), Mirzadeh & Mehrmanesh (2016), Rosemary et.al (2012), Mikit et.al (2012), Spenkar (2013)	media advertisement	
Navaei et.al (2015), Mosavi Esmaeilzadeh (2015), Hosseinzadeh et.al (2016), Ashouri (2016), Hassanzadeh & Danaei (2016), Mikit et.al (2012)	multimedia advertisement	
Navaei et.al (2015), Mirzadeh & Mehrmanesh (2016)	radio advertisement	
Navaei et.al (2015), Mosavi Esmaeilzdeh (2015), Ashouri (2016), Hassanzadeh & Danaei (2016)	cell phone advertisement	
Navaei et.al (2015), Hosseinzadeh et.al (2016), Mirzadeh & Mehrmanesh (2016)	creating TV advertisement	
Ahmadi (2013), Salimi Fard & Pourdehghan (2016), Hoang et.al (2013)	attraction of rewards	
siadat & Sakhi Jourashk (2014), Koumra et.al (2016)	being benevolent	

Darzaian (2016), Zabkouich & Louksten (2011), Abdolsalam et.al (2016)	coordination in speaking and behavior	
Ahmadi (2013), Salimi Fard & Pourdehghan(2016), Honari et.al (2015)	reality of messages	
Madhoushi (2013), Salimi Fard & Pourdehghan (2016), Houang et.al (2013)	graphic-designing	
Shahmohammadi (2011), Madhoushi (2013), Salimi Fard & Pourdehghan(2016)	psychological – geographical division	
Mirzadeh & Mehrmanesh (2016), Zabkouich & Lokesten (2011), Menroz et.al (2012), Choubhai et.al (2013), Mokan & Oltz (2014), Ayono et.al (2014)	advertisement on web	
Dehdashti Shahrokh (2009), Shahmohammadi (2011), Mirzadeh & Mehrmanesh (2016), Menroz et.al (2012), Choubhai et.al (2013), Mokan & Oltz (2014), Ayono et.al (2014), Henri et.al (2015), Yazghiro- Basteh (2015), Lin (2016)	internet & cell phone advertisement	
Shahmohammadi (2011), Sanei et.al (2016), Menroz et.al (2012), Chobai et.al (2013), Ayono et.al (2014), Henri et.al (2015)	selecting advertising media	
Shahmohammadi (2011), Sanei et.al (2016), Howang et.al (2013), Henri et.al (2015), Yazghirdo-Yasteh (2015), Martina Santa et.al (2017)	artistic aspect of advertisement	
Kashizadeh (2010), Hoshivar(2014), Pelpin et.al (2017)	gender of speaker in the advertisement	
Hosseinzadeh et.al (2016), oribeh (2016), Madra et.al (2017)	text repetition	
Shahmohammadi (2011), Hassanzadeh & Danaei (2016), Abri & Joumar (2014), Ayouno et.al (2014), Oribeh (2016), Pachman (2016), Bak (2016), Madra et.al (2017)	advertisement	
Samsam Shariat (2007), Sohrabi Ranani (2016), Daneshvar et.al (2016), Bak (2016), Kiji-Kali (2017)	factors relating to usage	environment

Fakhari et.al (2012), Mirzadeh & Mehrmanesh (2016), Howang et.al (2013), Bak (2016), kiji-Kali (2017)	factors relating to facilitation	
Dargi & Mohammadian (2010), Hamidizadeh (2013), Darzban (2016), Belman et.al (2012)	awareness of different status of the environment status	
Dargi & Mohammadian (2010), Mirzadeh & Mehrmanesh (2016)	awareness of access location of objective group to the product	
Dargi & Mohammadian (2010), Salimifard & Pourdehghan (2016)	awareness of norms	
Dargi & Mohammadian (2010), Hamidizadeh (2013), Darzban (2016), Belman et.al (2012), Koumra et.al (2016)	awareness of religions and religious values	
Dargi & Mohammadian (2010), Hassanzadeh & Danaei (2016)	awareness of method of evaluating services in relation to other services	
Hassazadeh & Danaei (2016), Radmiker (2015)	attention to security	
Samsam Shariat (2007), Mirzadeh & Mehrmanesh (2016), Belman et.al (2012)	economic beliefs	
Salimifard & Pourdehghan (2016), Pilay et.al (2012), Howang et.al (2013), Komra et.al (2016)	ethical beliefs	
Salimifard & Pourdehghan (2016)	Islamic teaching	
Samsam Shariat (2007), Daneshvar et.al (2016), Radmiker (2015)	strategic look to advertisement plan	
Dargi & Mohamadian (2010), Ansari & Rasei (2016), Abri & Joumar (2014)	market share	
Dargi & Mohamadian (2010), Darzain (2016)	obviating primary needs of audiences	

Salimi Fard & Pourdehghan (2016), Kim et.al (2012), Howang et.al (2013)	reference groups	
Sohrabi Ranin (2016), Fontini et.al (2009), Belman et.al (2012)	awareness of services	factors relating to services
Salimi Fard & Pourdehghan (2016), Abri & Joumar (2014), Radmiker (2015)	stating potential benefits of purchase ad services consumption	
Dargi & Mohamadian (2010), Hosseinzadeh et.al (2016), Fontini et.al (2009)	rate of services complexity	
Samsam Shariat (2007), Darzaian (2016), Fontini et.al (2009), Radmiker (2015)	display of expected experience	
Khashizadeh (2010), Darzain (2016)	rate of familiarity of consumer with services	
Ansari & Rasaei (2016), Belman et.al (2012)	market researches	
Hosseinzadeh et.al (2016), Mirzadeh & Mehrmanesh (2016)	record of exploiting services	
Dargi & Mohammadian (2010), Ansari & Rasaei (2016), Lousi (2012), Park (2015), Marion et.al (2017)	being unique	
Dehdashti Shahrokh (2009), Daneshvar et.al (2016), Radmiker (2015), Jankoski et.al (2016)	price, benefits	
Daneshvar et.al (2016), Fontini et.al (2009), Jankoski et.al (2016)	relation with customer	
Samsam Shariat (2007), Daneshvar et.al (201), Fontini et.al (2009), Jankoski (2016)	supporting customer	
Dehdashti Shahrokh (2009), Ansari & Rasaei (2016), Fontini et.al (2009), Radmiker (2015), Jankoski (2016)	feedback	

2- Research plan

The present study is of descriptive applied type which has been performed with survey method. The statistical population of this study includes all managers and experts of Eghtesad

Novin Bank. The statistical population of the present study contains all people who are exposed to advertisements of Eghtesad Novin Bank across the country. For selecting samples, Cochran formula is used for unlimited societies that based on this formula, the samples number were determined 400 persons for each province. The sample people were selected based on the following sampling method. At first, using stratified sampling method , the samples were randomly selected from five geographical regions of the country (north, south, east, west and center) of nine provinces (Mazandaran, Khorasan Razavi, Zanjan, Chahar Mahal Bakhtiari , Kerman, East Azarbaijan , Kermanshah, Ilam and Tehran).

3- Variables and research model

In this study, fuzzy Delphi technique has been used for identifying main indicators. For ranking the final criteria and sub standards, multi-criteria decision making models and AHP model have been used with fuzzy approach. For analysis of obtained data, Excel software and coding in the environment of this software have been used. The results obtained from data analysis have been provided in the present chapter.

3-1 Identifying criteria and sub standards of factors effective on advertisement effectiveness

At the first step, the primary indicators were identified and selected. In the first stage, by reviewing the literature and performed studies, a large number of indicators were extracted from the literature. For screening the indicators and identifying the final indicators, fuzzy Delphi approach has been used. The expert views have been collected in the field of rate of significance of main indicators. Fuzzy average and defuzzification output of rates relating to indicators are presented in table 2. The rate of defuzzification greater than 7 is acceptable and every indicator which has the score higher than 7 is confirmed (Wu& Fang, 2011).

Table 2- fuzzy average and fuzzy screening of indicators of factors effective on advertisement effectiveness

Result	Crisp	mean	U	M	L	
acceptance	7.15	(2, 7.97, 9)	9	7.97	2	Q01
acceptance	7.84	(6, 8.01, 9)	9	8.01	6	Q02

Result	Crisp	mean	U	M	L	
acceptance	7.92	(6, 8.13, 9)	9	8.13	6	Q03
acceptance	7.39	(4, 7.83, 9)	9	7.83	4	Q04
acceptance	7.70	(6, 7.8, 9)	9	7.80	6	Q05
rejection	6.68	(2, 7.28, 9)	9	7.28	2	Q06
acceptance	7.71	(6, 7.82, 9)	9	7.82	6	Q07
acceptance	7.66	(6, 7.75, 9)	9	7.75	6	Q08
acceptance	7.02	(2, 7.78, 9)	9	7.78	2	Q09
acceptance	7.27	(2, 8.15, 9)	9	8.15	2	Q10
acceptance	7.09	(2, 7.88, 9)	9	7.88	2	Q11
rejection	6.85	(2, 7.52, 9)	9	7.52	2	Q12
acceptance	7.43	(4, 7.9, 9)	9	7.90	4	Q13
rejection	6.71	(3, 7.07, 9)	9	7.07	3	Q14
acceptance	7.17	(2, 8.01, 9)	9	8.01	2	Q15
acceptance	7.13	(2, 7.95, 9)	9	7.95	2	Q16
rejection	6.85	(4, 7.02, 9)	9	7.02	4	Q17
acceptance	7.43	(4, 7.89, 9)	9	7.89	4	Q18
acceptance	7.70	(6, 7.8, 9)	9	7.80	6	Q19
rejection	6.80	(3, 7.21, 9)	9	7.21	3	Q20
acceptance	7.75	(6, 7.88, 9)	9	7.88	6	Q21
rejection	6.62	(3, 6.93, 9)	9	6.93	3	Q22
acceptance	7.61	(6, 7.66, 9)	9	7.66	6	Q23
acceptance	7.22	(2, 8.08, 9)	9	8.08	2	Q24
acceptance	7.04	(2, 7.81, 9)	9	7.81	2	Q25
acceptance	7.71	(6, 7.82, 9)	9	7.82	6	Q26
rejection	6.65	(3, 6.97, 9)	9	6.97	3	Q27
acceptance	7.98	(6, 8.22, 9)	9	8.22	6	Q28
acceptance	7.70	(6, 7.79, 9)	9	7.79	6	Q29
rejection	6.54	(3, 6.81, 9)	9	6.81	3	Q30
acceptance	7.19	(4, 7.53, 9)	9	7.53	4	Q31
acceptance	7.79	(5, 8.18, 9)	9	8.18	5	Q32

Result	Crisp	mean	U	M	L	
acceptance	7.53	(4, 8.05, 9)	9	8.05	4	Q33
acceptance	7.05	(2, 7.82, 9)	9	7.82	2	Q34
acceptance	7.43	(4, 7.89, 9)	9	7.89	4	Q35
rejection	6.92	(3, 7.38, 9)	9	7.38	3	Q36
acceptance	7.02	(2, 7.78, 9)	9	7.78	2	Q37
rejection	6.91	(3, 7.37, 9)	9	7.37	3	Q38
rejection	6.96	(2, 7.68, 9)	9	7.68	2	Q39
rejection	6.99	(2, 7.74, 9)	9	7.74	2	Q40
acceptance	7.42	(4, 7.88, 9)	9	7.88	4	Q41
acceptance	7.10	(2, 7.9, 9)	9	7.90	2	Q42
rejection	6.59	(2, 7.13, 9)	9	7.13	2	Q43
rejection	6.55	(2, 7.08, 9)	9	7.08	2	Q44
acceptance	7.77	(6, 7.9, 9)	9	7.90	6	Q45
rejection	6.91	(2, 7.62, 9)	9	7.62	2	Q46
acceptance	7.09	(2, 7.88, 9)	9	7.88	2	Q47
acceptance	7.68	(6, 7.78, 9)	9	7.78	6	Q48
acceptance	8.19	(6, 8.53, 9)	9	8.53	6	Q49
rejection	6.88	(2, 7.57, 9)	9	7.57	2	Q50
acceptance	7.75	(5, 8.12, 9)	9	8.12	5	Q51
acceptance	7.03	(2, 7.8, 9)	9	7.80	2	Q52
rejection	6.96	(4, 7.19, 9)	9	7.19	4	Q53
acceptance	7.04	(2, 7.8, 9)	9	7.80	2	Q54
rejection	6.96	(3, 7.45, 9)	9	7.45	3	Q55
acceptance	7.02	(2, 7.78, 9)	9	7.78	2	Q56
acceptance	7.92	(5, 8.38, 9)	9	8.38	5	Q57
acceptance	7.00	(2, 7.76, 9)	9	7.76	2	Q58
rejection	6.90	(1, 7.86, 9)	9	7.86	1	Q59
acceptance	7.10	(4, 7.4, 9)	9	7.40	4	Q60
acceptance	7.86	(6, 8.04, 9)	9	8.04	6	Q61
rejection	6.97	(2, 7.71, 9)	9	7.71	2	Q62

Result	Crisp	mean	U	M	L	
acceptance	7.02	(2, 7.79, 9)	9	7.79	2	Q63
acceptance	7.73	(6, 7.84, 9)	9	7.84	6	Q64
acceptance	7.01	(2, 7.76, 9)	9	7.76	2	Q65
rejection	6.65	(3, 6.98, 9)	9	6.98	3	Q66
acceptance	7.01	(2, 7.77, 9)	9	7.77	2	Q67
acceptance	7.90	(6, 8.11, 9)	9	8.11	6	Q68
acceptance	7.55	(5, 7.83, 9)	9	7.83	5	Q69
acceptance	8.02	(6, 8.28, 9)	9	8.28	6	Q70
acceptance	7.95	(6, 8.17, 9)	9	8.17	6	Q71
acceptance	7.03	(2, 7.79, 9)	9	7.79	2	Q72
acceptance	7.65	(5, 7.98, 9)	9	7.98	5	Q73
acceptance	7.00	(1, 8, 9)	9	8.00	1	Q74
acceptance	7.04	(2, 7.8, 9)	9	7.80	2	Q75
acceptance	7.03	(2, 7.8, 9)	9	7.80	2	Q76
acceptance	7.30	(4, 7.7, 9)	9	7.70	4	Q77
rejection	6.88	(2, 7.57, 9)	9	7.57	2	Q78
acceptance	7.63	(4, 8.19, 9)	9	8.19	4	Q79
acceptance	7.65	(5, 7.98, 9)	9	7.98	5	Q80
rejection	6.59	(3, 6.88, 9)	9	6.88	3	Q81
acceptance	7.09	(2, 7.88, 9)	9	7.88	2	Q82
rejection	6.83	(2, 7.5, 9)	9	7.50	2	Q83
acceptance	7.12	(2, 7.93, 9)	9	7.93	2	Q84

From the 84 indicators which were assessed, 24 indicators which have acquired the average score less than 7 are omitted. So, 60 indicators are finally confirmed (Adelazaret.al, 1995).

The remained indicators will be finally examined in the format of a 60-question questionnaire with analytical hierarchy process.

3-2 determining the priority of factors effective on effectiveness of advertisement using fuzzy AHP technique

For determining the priority of major criteria, fuzzy analytical hierarchy (FAHP) has been used. The analysis process is as follows.

1- Dual comparison of major criteria based on objective and determining the weight of major criteria

2- Dual comparison of sub standards of each criterion and determining sub standards of each cluster

3- Multiplication of the weight of sub standards in the weight of respective criterion and determining the final weight of sub standards

For dual comparison of elements, 9-degree Saaty scale has been used. 9-degree Saaty scale has been provided by Thomas Saaty , codifier of analytical hierarchy theory. Also, in this study fuzzy approach has been used for making the values quantitative. So, Saaty fuzzy range has been used.

Table 3- fuzzy range equal to 9-degree Saaty scale in AHP technique. Habibi et.al , 2014, 77

reversed fuzzy equivalent	fuzzy equivalent	the word term of comparison status of i to j
(1,1,1)	(1, 1, 1)	Preferred Equally
$\left(\frac{1}{3},\frac{1}{2},1\right)$	(1, 2, 3)	intermediate
$\left(\frac{1}{4},\frac{1}{3},\frac{1}{2}\right)$	(2, 3, 4)	Preferred moderately
$\left(\frac{1}{5},\frac{1}{4},\frac{1}{3}\right)$	(3, 4, 5)	intermediate
$\left(\frac{1}{6},\frac{1}{5},\frac{1}{4}\right)$	(4, 5, 6)	Preferred Strongly
$\left(\frac{1}{7},\frac{1}{6},\frac{1}{5}\right)$	(5, 6, 7)	intermediate
$\left(\frac{1}{8},\frac{1}{7},\frac{1}{6}\right)$	(6, 7, 8)	very strongly Preferred
$\left(\frac{1}{9},\frac{1}{8},\frac{1}{7}\right)$	(7, 8, 9)	intermediate
$\left(\frac{1}{9},\frac{1}{9},\frac{1}{9}\right)$	(9, 9, 9)	Extremely Preferred

(source in papersL Khorshid & Zabihi, Lee my H.I; Wen-Chin Chen, Ching-Jan, Chang, 2008, 101)

The pattern of criteria and sub standards has been drawn in the figure using AHP technique. Also, the criteria and sub standards of the study have been named with numeric index so that the research flow is easily detectable.

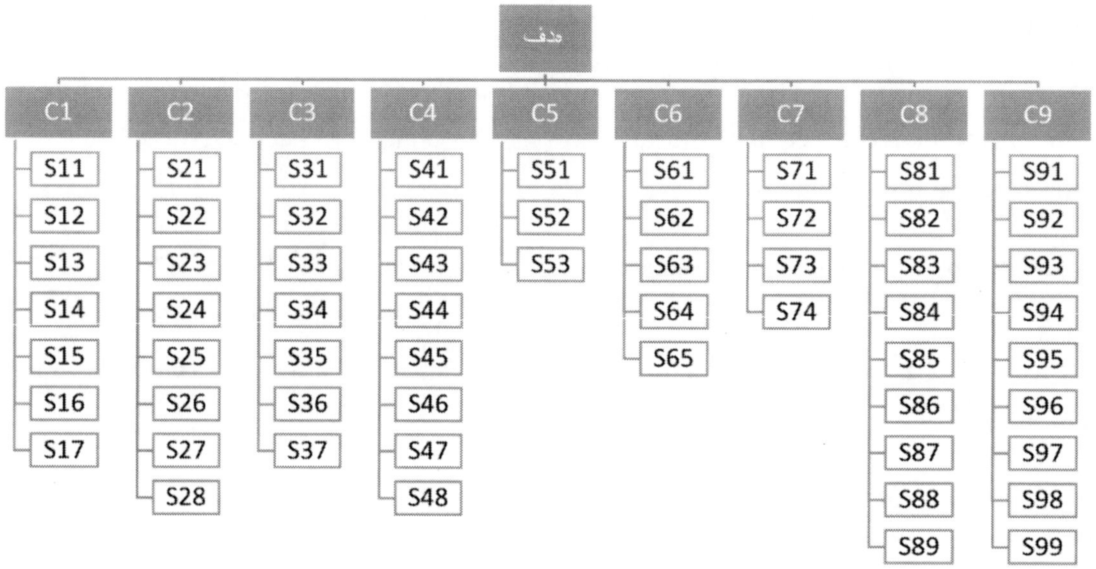

Figure 1- hierarchical model of criteria and sub standards of factor effective on advertisement effectiveness

1- Determining the priority of major criteria based on objective

In the first step, the major criteria based on the objective have been compared dually. Dual comparison is very simple and all elements of each cluster should be compared two by two. Therefore, if there is n element in a cluster, $\frac{n(n-1)}{2}$ comparisons will be performed. Since there are four criteria, so the number of performed comparisons is equal to:

$$\frac{n(n-1)}{2} = \frac{9(9-1)}{2} = 36$$

So, 36 dual comparisons have been performed from group perspective of experts. The experts view has been made quantitative using fuzzy scale. Collecting experts view has been collected with 9-degree Saaty range. So, 6 dual scales of four major four criteria have been performed based on the experts view. Using geometric average the experts views have been aggregated. Fuzzy aggregation will be calculated as follows:

Equation 1

$$F_{AGR} = (\prod(l), \prod(m), \prod(u))$$

After matrix formation of the obtained dual comparisons, the special vector has been calculated. At first, the fuzzy extension of each row is calculated. Every entry of dual comparison matrix $\widetilde{X}$

Is displayed as $\tilde{x}_{ij}$.

The fuzzy extension of every row is also displayed with symbol $\tilde{S}_i$. Therefore, the fuzzy extension of each row is calculated as follows:

Equation 2

$$\tilde{S}_i = \sum_{j=1}^{n} x_{ij}$$

Then, the fuzzy extension of every row elements will be as follows:

Fuzzy extension of row 1 (1,1,1) ⊕ (1,27,1,47,1,75) ⊕ (0.47, 0.55, 0.66) ⊕ (1.52, 1.81, 2.14) ⊕ (0.93, 1.14, 1.39) ⊕ (1.26, 1.61, 2.04) ⊕(1.25, 1.51, 1.84) ⊕(0.78, 0.97, 1.24) ⊕ (0.59, 0.75, 0.96) ⊕ (9.08, 10.8, 13.02)

Fuzzy extension of row 2 (0.57, 0.68, 0.79)⊕(1, 1, 1)⊕(0.58, 0.7, 0.89)⊕(1.84, 2.28, 2.75)⊕(0.68, 0.82, 1.02)⊕(1.42, 1.81, 2.34)⊕(1.64, 2.04, 2.53)⊕(0.71, 0.88, 1.12)⊕(0.73,
fuzzy extension of row 3 (1.52, 1.82, 2.11)⊕(1.13,)0.9, 1.1)⊕ = (9.17, 11.11, 13.55
1.42, 1.72)⊕(1, 1, 1)⊕(1.89, 2.34, 2.88)⊕(0.7, 0.86, 1.09)⊕(1.45, 1.78, 2.22)⊕(1.55, 1.88, 2.29)⊕(0.57, 0.71, 0.93)⊕(0.67, 0.8, 0.96)⊕ = (10.47, 12.61, 15.2)

Fuzzy extension of row 4 (0.47, 0.55, 0.66)⊕(0.36, 0.44, 0.54)⊕(0.35, 0.43, 0.53)⊕(1, 1, 1)⊕(0.97, 1.21, 1.51)⊕(0.96, 1.33, 1.8)⊕(2.14, 2.7, 3.37)⊕(0.94, 1.21, 1.56)⊕(1.3, 1.69, 2.16)⊕ = (8.49, 10.57, 13.14)

Fuzzy extension of row 5 (0.72, 0.88, 1.07) ⊕ (0.98, 1.22, 1.48)⊕(0.92, 1.17, 1.43)⊕(0.66, 0.82, 1.03)⊕(1, 1, 1)⊕(0.67, 0.83, 1.04)⊕(0.23, 0.29, 0.36)⊕(0.28, 0.32, 0.39)⊕(0.3, 0.38, 0.48)⊕ = (5.76, 6.91, 8.29)

Fuzzy extension of row 6 (0.49, 0.62, 0.79)⊕(0.43, 0.55, 0.7)⊕(0.45, 0.56, 0.69)⊕(0.55, 0.75, 1.04)⊕(0.96, 1.2, 1.48)⊕(1, 1, 1)⊕(0.34, 0.41, 0.51)⊕(0.27, 0.34, 0.44)⊕(0.24, 0.31, 0.41)⊕ = (4.74, 5.75, 7.08)

Fuzzy extension of row 7 (0.54, 0.66, 0.8)⊕(0.4, 0.49, 0.61)⊕(0.44, 0.53, 0.64)⊕(0.3, 0.37, 0.47)⊕(2.81, 3.5, 4.3)⊕(1.95, 2.43, 2.96)⊕(1, 1, 1)⊕(0.34, 0.42, 0.54)⊕(0.29, 0.36, 0.47)⊕ = (8.06, 9.76, 11.78)

Fuzzy extension of row 8 (0.8, 1.03, 1.28)⊕(0.89, 1.14, 1.41)⊕(1.08, 1.41, 1.76)⊕(0.64, 0.82, 1.06)⊕(2.55, 3.1, 3.63)⊕(2.25, 2.97, 3.72)⊕(1.86, 2.39, 2.94)⊕(1, 1, 1)⊕(0.58, 0.69, 0.83)⊕ = (11.66, 14.56, 17.64)

Fuzzy extension of row 9 (1.04, 1.33, 1.69)⊕(0.91, 1.11, 1.37)⊕(1.04, 1.25, 1.5)⊕(0.46, 0.59, 0.77)⊕(2.08, 2.63, 3.3)⊕(2.42, 3.23, 4.1)⊕(2.15, 2.74, 3.43)⊕(1.2, 1.46, 1.72)⊕(1, 1,)1)⊕ = (12.29, 15.35, 18.88

Then, the fuzzy sum of total priorities column elements is calculated

Equation 3

$$\sum \tilde{S}_i = \sum_{i=1}^{n}\sum_{j=1}^{n} x_{ij}$$

The sum of elements of the priorities column of major criteria will be as follows:

$$\sum_{i=1}^{4}\sum_{j=1}^{4} M_g^j = (79.72, 97.41, 118.59)$$

For normalization of each criterion priorities, that criterion total rates sum should be divided into sum of all priorities (column elements), since the rates are fuzzy, so fuzzy sum of every row is multiplied in summation reverse. The summation reverse should be calculated.

Equation 4

$$if\ \tilde{F} = (l, m, u)\ then\ \tilde{F}^{-1} = \left(\frac{1}{u}, \frac{1}{m}, \frac{1}{l}\right)$$

Then, based on the equation 4, we will have

$$\left(\sum_{i=1}^{n}\sum_{j=1}^{n} M_g^j\right)^{-1} = (0.01, 0.01, 0.01)$$

$$P_k = \sum_{i=1}^{n} M * \left(\sum_{i=1}^{n}\sum_{j=1}^{n} M_g^j\right)^{-1}$$

So, the results obtained from normalization of obtained rates will be as follows:

$\tilde{W}_{C1}$ = (0.08, 0.11, 0.16)

$\tilde{W}_{C2}$ = (0.08, 0.11, 0.17)

$\tilde{W}_{C3}$ = (0.09, 0.13, 0.19)

$\tilde{W}_{C4}$ = (0.07, 0.11, 0.16)

$\tilde{W}_{C5}$ = (0.05, 0.07, 0.1)

$\widetilde{W}_{C6} =$ (0.04, 0.06, 0.09)

$\widetilde{W}_{C7} =$ (0.07, 0.1, 0.15)

$\widetilde{W}_{C8} =$ (0.1, 0.15, 0.22)

$\widetilde{W}_{C9} =$ (0.1, 0.16, 0.24)

Each of the obtained rates is fuzzy and normalized weight relating to major criteria. There are various methods for defuzzifcation of the obtained rates. In this study, Bojadziev proposed method has been used for defuzzification.

Table 4- defuzzification and computation of final and normal weight of criteria of factors effective on advertisement effectiveness

Normal	Deffuzy	X3max	X2max	X1max	Crisp
0/111	0/117	0/113	0/115	0/117	C1
0/114	0/120	0/115	0/118	0/120	C2
0/103	0/137	0/133	0/135	0/137	C3
0/108	0/113	0/112	0/113	0/113	C4
0/070	0/073	0/072	0/073	0/073	C5
0/060	0/063	0/062	0/063	0/063	C6
0.101	0/107	0/103	0/105	0/107	C7
0.149	0/157	0/153	0/155	0/157	C8
0.158	0/167	0/163	0/165	0/167	C9

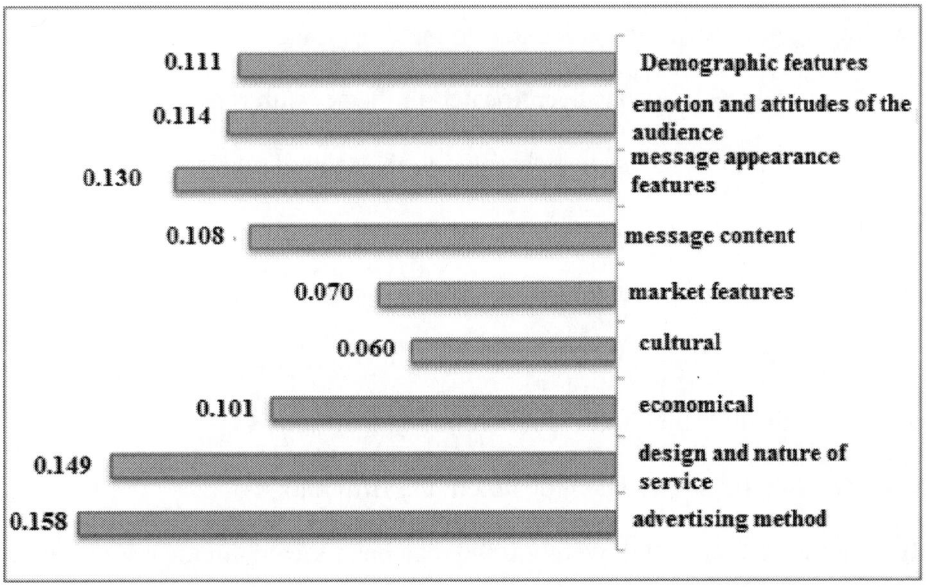

Figure 2- graphical display of significance of criteria of factors effective on advertisement effectiveness

Based on the obtained special vector:

- Demographic features with weight 0.111 have been located in rank five

- Individual feelings and attitude of the addressee with weight 0.114 has been located in rank four.

- The message surface features with weight 0.13 has been located in rank three.

- The message content with weight 0.108 has been located in rank six

- The market feature with weight 0.07 has been located in rank eight.

- Cultural with weight 0.06 has been located in rank nine.

- Economic with weight 0.10 has been located in rank seven.

- Plan and nature of services with weight 0.149 has been located in rank second.

- Advertisement methods with weight 0.158 has been located in rank first.

The inconsistency rate of performed comparisons has been obtained 0.030 which is smaller than 0.1 and so the performed comparisons is sufficed.

2- Comparison and determining priority of sub standards

At first, the experts' views have been collected in the dual comparison of sub standards. Then, the performed calculations for fuzzification of the experts views average have been provided for determining the priority of sub standards.

Priority of sub standards of demographic features

The audience job with weight 0.151 has been located in the third rank.

The audience income with weight 0.093 has been located in the seventh rank.

The audience religion with weight 0.16 has been located in the second rank.

Social class with weight 0.141 has been located in the fifth rank.

Awareness of gender differences with weight 0.173 has been located in the first rank.

The individual knowledge level with weight 0.135 has been located in the sixth rank.

Priority of sub standards of the audience individual feeling and attitude

Positive mentality with weight 0.124 has been located in the fifth rank.

Feeling of honesty in advertisements with weight 0.089 has been located in the eighth rank.

Threatening feeling with weight 0.14 has been located in the third rank.

Mental image with weight 0.102 has been located in the seventh rank.

Attitude towards brand with weight 0.121 has been located in the sixth rank.

Convincing audiences with weight 0.149 has been located in the first rank.

Attitude towards advertisement with weight 0.143 has been located in the second rank.

The individual perception of accuracy of advertisement with weight 0.133 has been located in the fourth rank.

Priority of sub standards of the message surface features

Then, based on the results of performed dual comparisons:

Beautiful and appropriate combination of colors with weight 0.143 has been located in the fourth rank.

Attention to reduction of the user waiting time with weight 0.159 has been located in the second rank.

Graphics- design with weight 0.131 has been located in the sixth rank.

Creativity in the message with weight 0.137 has been located in the fifth rank.

Emotional music with weight 0.099 has been located in the seventh rank.

Text repetition with weight 0.153 has been located in the third rank.

Access to required and sufficient information with weight 0.178 has been located in the first rank.

Determining the priority of sub standards of the message content

Benevolent with weight 0.093 has been located in the eighth rank.

Coordination in speech and behavior with weight 0.125 has been located in the fourth rank.

Reality of the messages with weight 0.119 has been located in the sixth rank.

Using women and children images with weight 0.131 has been located in the third rank.

Gender of speaker in the advertisement with weight 0.106 has been located in the seventh rank.

Using trustful and famous people with weight 0.16 has been located in the first rank.

Loveliness of famous people for the observer with weight 0.122 has been located in the fifth rank.

Familiarity of the famous person with the observer with weight 0.144 has been located in the second rank.

Priority of sub standards of the market features

Then, based on the results of the performed dual comparisons:

Factors relating to its application in the market with weight 0.326 have been located in the second rank.

Awareness of different aspects of status and environment with weight 0.283 has been located in the third rank.

Awareness of evaluation method of service receiver with weight 0.391 has been located in the first rank.

Priority of cultural sub standards

Then, based on the results of performed dual comparisons:

Awareness of norms with weight 0.199 has been located in the third rank.

Awareness of religion and religious values with weight 0.238 has been located in the first rank.

Ethical beliefs with weight 0.18 have been located in the fourth rank.

Obviating primary needs of the audiences with weight 0.164 has been located in the fifth rank.

Reference groups with weight 0.219 have been located in the second rank.

Determining the priority of economic sub standards

Then, based on the results of the performed dual comparison:

Economic beliefs with weight 0.261 have been located in the second rank.

Strategic look at an advertising program with weight 0.215 has been located in the fourth rank.

Market share with weight 0.274 has been located in the first rank.

Attention to economic security with weight 0.251 has been located in the third rank.

Priority of the plan and services nature sub standard

Awareness of services with weight 0.104 has been located in seventh rank.

Expressing potential benefits of purchase and services consumption with weight 0.117 has been located in the third rank.

Services complexity rate with weight 0.089 has been located in the ninth rank.

Uniqueness with weight 0.108 has been located in the sixth rank.

Rate of familiarity of the services consumer with weight 0.12 has been located in the second rank.

Price and allowances with weight 0.114 has been located in the fourth rank.

Relation with the customer with weight 0.111 has been located in the fifth rank.

Displaying the expected experience with weight 0.098 has been located in the eighth rank.

Supporting customers with weight 0.139 has been located in the first rank.

Priority of sub standards of the advertising methods

Beginning the advertisement with a challenge with weight 0.079 has been located in the ninth rank.

Beginning an advertisement with a statement and explanation with weight 0.107 has been located in the fifth rank.

Using stimuli with diversions more than usual and expected with weight 0.097 has been located in the seventh rank.

The advertisement message text being predicating with weight 0.126 has been located in the third rank.

Using diverse methods with weight 0.088 has been located in the eighth rank.

Media advertisement with weight 0.113 has been located in the fourth rank.

Dimensions of TV advertisement with weight 0.151 have been located in the first rank.

Psychological, geographical division with weight 0.107 has been located in the fifth rank.

Advertising media selection with weight 0.132 has been located in the second rank.

Determining the final priority of major criteria with FAHP technique

For determining final priority of major criteria using FAHP technique, the weights relating to major criteria (W1) and weight of indicators based on each criteria (W2) are needed. The results of comparison of the research sub criteria relating to them forms matrix W2. For determining the final priority of indicators with AHP technique, it is sufficient that the weight of indicators based on every criterion (W2) is multiplied in the major criteria (W1). Any of these matrices has been calculated in the previous steps. Based on the obtained output, it is obvious that the market share item, one of economic factors with weight 0. 277 is the most important indicator in the existing indicators. Item of awareness of evaluation method of service receiver in relation to other services with weight of 0.0273 has the second priority. Item of economic believes, one of economic factors, has the third priority.

So, the research model is as follows:

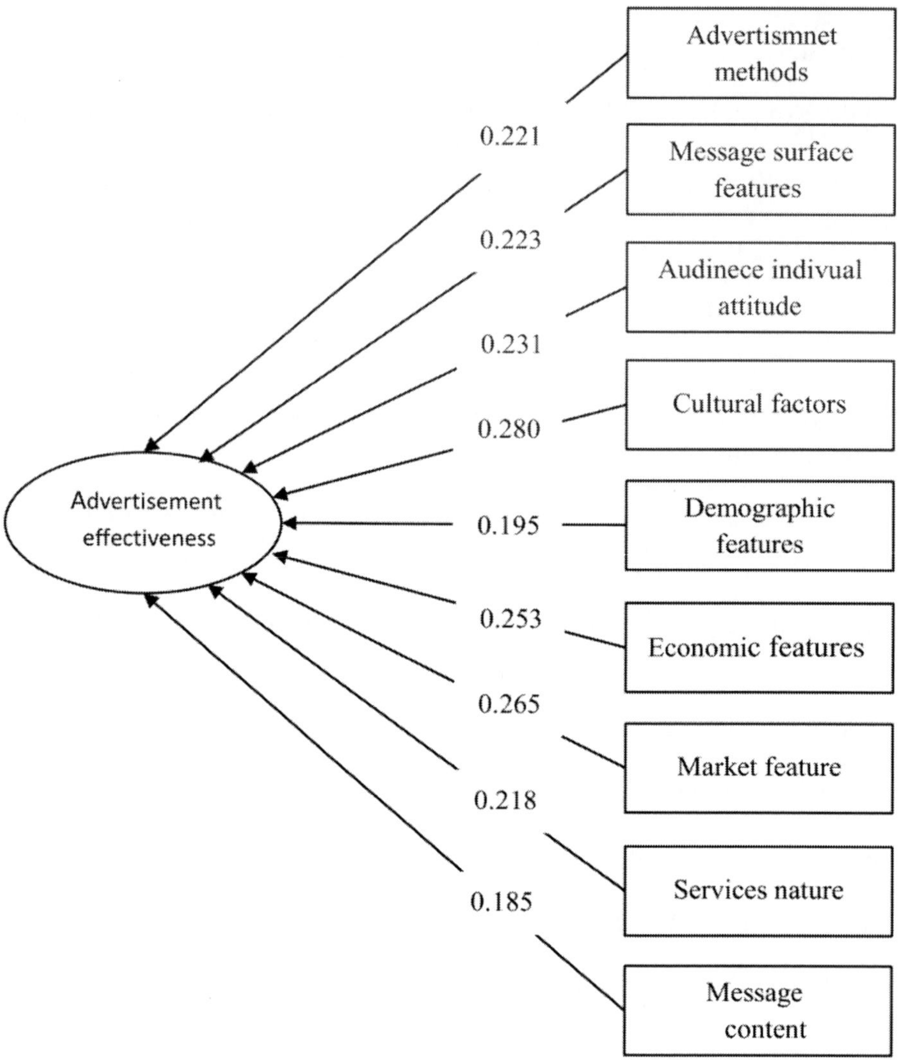

4- Conclusion:

In this study, identification and ranking of factors effective on designing the native model of advertisement in Eghtesad Novin bank has been addressed. In this first step, the primary indicators were identified and selected. In the first stage, by reviewing the literature and performed studies, a great number of indicators were extracted from the literature. For screening the indicators and identifying final indicators, fuzzy Delphi approach has been used. The experts' views have been collected in the field of significant rate of major indicators. From 84 indicators which were assessed, 24 indicators which have acquired score less than 7 are omitted. So, 60 indicators were finally confirmed. Then, regarding the performed calculations, the final weight of any of the model indicators have been calculated with fuzzy AHP. Based on the obtained output, it is obvious that the market share of economic factors with weight of 0.027 is the most important indicator in the existing indicators. The item of awareness of evaluating services in relation to other services is one of market features factors with weight 0.0273 which has second priority. The item of economic beliefs with weight 0.0264 is in the third priority.

Today, considering extension of competition in the international level and formation of world competition, all countries of the world may be successful in this arena to coordinate them with this trend and create more value for their customers by services and products they provide. In the today modern world, almost all people are exposed to advertisements and other communicational activities. Effectiveness of advertisement is one of important issues in the advertisement domain. In this regard, Mansouri Moyed (2005) showed that attention attraction, commercial advertisements of Maskan Bank are effective but in respect of other variables, that is, creating interest, stimulation of tendency and measurement to purchase, commercial advertisements of Maskan Bank are not effective. Samsam Shariat (2007) showed that TV channels and cartoon animations , attractive package, colors relating to the product and innovation, noticing safety feeling and associated symbols, informing of quality, background and price, dialects and common phrases, diversity in the message content, in TV, broadcasting from 18-23 and before beginning programs (in media) have been effective factors in effectiveness of commercial effectiveness from people viewpoint. Mosavi Esmaeilzadeh (2015) showed that oral advertisements (in 4 dimensions) and after that TV advertisement (just in 2 dimensions) are effective. Also, radio, internet and billboard advertisement are not effective. And also Friedman test was used for ranking advertisement tools. Salimi Fard & Pourdehghan (2016) showed that among factors effective on attitude to cell phone advertisement, components of general attitude towards advertisement, knowledge level and value have respectively the greatest significance and components of time of presentation and reference groups with the least significance in effecting the individuals attitude towards cell phone advertisement. Ansari & Rasaei (2016) showed that the creativity of advertisement message , market research, competition, market share, uniqueness and relation with customer are effective on the effectiveness of brand advertisement. Louis (2012) showed that advertising in the urban public spaces will have a great impact in the consumers' mentality. Since some advertisements are transitory and are often displayed in a short period of time and just in several spaces, but advertisement in public spaces exist everywhere and is similar to advertising through print and media, but since they have lower cost for production and distribution, have lower cost in this respect. Zabckowich & Loukston (2011) showed that in necktie advertisements use more of sensory requests and information signs. Findings showed that in western and European countries, people establish more contact due to globalization increase. They declared that content of printed advertisements, includes visual and text advertisements which are effective in establishing relation with many audiences. Also, ethnic identity as an important cultural variable increases attention to consumer purchase behavior. Besides that, they found out that the image of people, models, ceremonies and managers constitute an important part of images and background image plays an important role in effectiveness of visual factors in advertisement. Mac Ki- Nesbiat et.al (2011) showed that younger adults remind emotional messages specially negative ones better than others and older adults more prefer positive and logical messages to negative and emotional ones, but attitude towards advertisement among younger adults is not different. Also, they found that older adults have the most positive attitude towards advertisements.

Perzouti (2015) showed that advertising models are different in various ages and each age is consistent with a special model and using various and consistent models with people conditions may make the advertisement more effective. Abri & Joumar (2014) showed that advertising messages are effective on the consumer behavior and the advertiser may use various media for communicating consumer and media schedule is effective on the advertisement success. Attention to environmental factors where the customer lives are important factors in the advertisement success. Madra et.al (2017) showed that the advertisement text, text repetition impact the advertisement effectiveness. Maryion et.al (2017) showed that simultaneous use of several media like TV and internet may increase effectiveness of the advertisement and considering two variables of gender and multimedia create a better individual attitude towards the message effectiveness.

In ranking the identified factors, advertisement methods have assigned the first rank among factors. It is suggested that in designing advertisement and selecting advertising instruments, the dimensions of TV advertising, selecting median and predicative message text should be noticed. Using advertising instrument like internet advertisement, adverting on public vehicles like buses and metro due to great number of people who use them can increase stability of the advertisements in the customers mind.

In codifying and designing bank advertisements, the design and services nature should be noticed. It is proposed that supporting customer, familiarity of consumer with services are considered and during the advertisement process, it is required that the potential benefits of purchase and services consumption are expressed.

In designing the advertisement and content of bank advertising message, it is required to use famous and trustful people for introducing the services. Also, children as future customers and audiences should be noticed since they are future potentials. Total advertisements provided by the bank through various media should move in a unit line so that the audience can establish a logical relation between the observed advertisements in various media that is total advertising efforts lead to synergy of advertisement effectiveness. This issue won't be acquired unless through a media planning and efficient advertisement.

In advertisements, the message surface features should be noticed. It is required that in advertisement, access to required and enough information for the customers, attention to reduction of expectation time of user for observing the advertisement and suitable repetition of the text are considered. Providing large information in an advertisement which has a short time and providing information that require special mathematical analysis should be avoided.

References:

Hosseinabadi, Maryam, Mahmoud Nematian, Mohammad Besharati, investigating the relation between mixed elements of marketing and customers satisfaction with Tehran

Sarmayeh bank branches, management and entrepreneurship studies, fall 2016, 2nd period, No. 3 (page 162-169)

Dargi Parviz, 2008, subjects and topics of marketing management with Iran market attitude, Tehran, Rasa publication

Sarmad, Zohreh, Bazargan Abbas, Hejazi, Elaheh, 2014, research methods in behavioral sciences, Tehran, Agah. 27-13

Salimi Fard, Khodakaram, Pourdehghan, Adel, 2016, identifying and ranking factors effective on attitude towards cell phone advertisement using AHP, Brand management, period 3, no. 6, summer 2016, page 133-160

Seyed Akbar Manjgani, Vahideh Tabasi Lotfabadi, 2015, factors effective on effectiveness of advertisement on financial stoop of legal clients of Mashhad City with AHP approach, finance journal, fall 2015, No. 75 (page 121-146)

Sanayei, Ali, Majid Mohammad Shafiei, Maedeh Amini Valashani, 2016, effectiveness of bank advertisement in customer attraction, commercial management searches, spring and summer 2016, No. 15 (page 185-210)

Mahmoudian, Mahmoud, Rabiei, Ali, Jamili, Bita. 2011, evaluation of effectiveness of Parsian bank advertisement and identifying the most important factors in increasing its effectiveness in Tehran city, modern marketing studies, summer 2011, No. 2 (page 17-40).

Manjagani, Seyed Akbar, Tabasi Lotfabadi, Vahideh, 2015, factors effective on advertisement effectiveness on financial stoop of legal clinets of Mashhad city with AHP approach, finance journal, fall 2015, No. 75, 1-121

Venous, Davar, Safaeian, Mitra, (2005), marketing bank services in Iran and bank services features, Tehran, Negah Danesh, 4th edition

Angulo, N. (2016). 10 digital marketing trends to watch in 2016 and beyond. Retrieved from: http://www.marketingdive.com/news/10-digital-marketing-trends-to-watch-in-2016-./and-beyond/411457

Bell Raoul, Buchner, Axel.(2018). Positive Effects of Disruptive Advertising on Consumer Preferences, Journal of Interactive Marketing 41 (2018) 1–13

Corvi. Elisabetta.(2010).The effectiveness of advertising: a literature review, 10th Global Conference on Business & Economics, ISBN : 978-0-9830452-1-2

Cramphorn Michael F.,(2011), Forum Gender effects in advertising, International Journal of .Market Research Society Vol. 53 Issue 2

Cronin, J. J. & Fox , G. L. (2010). "The implications of third party customer complaining for .advertising efforts". Journal of advertising. 39(2): 21-23

Decrop, A. (2007). The influence of message format on the effectiveness of print advertisements for tourism destinations. International Journal of Advertising, 26(4), 505–.525

Denis .A. Shakhov and ,Anna A. Panasenko.(2012).Evaluating Effectiveness of Bank Advertising in the Internet: Theory and Practice, World Applied Sciences Journal 18 (Special Issue of Economics): 83-90, 2012 ISSN 1818-4952

Denis 1,2 A. Shakhov and 3Anna A. Panasenko.(2012).Evaluating Effectiveness of Bank Advertising in the Internet: Theory and Practice, World Applied Sciences Journal 18 (Special Issue of Economics): 83-90, 2012

Durdana Ozretic-Dosen Ines Zizak , (2015),"Measuring the quality of banking services targeting student population", EuroMed Journal of Business, Vol. 10 Iss 1 pp. 98 - 117

Eckler, P., and Bolls, P. (2011). Spreading the Virus: Emotional Tone of Viral Advertising and Its Effect on Forwarding Intention and Attitudes. Journal of Interactive Advertising, .11(2): 1-11

Eric W.T. Ngai Ka-leung Karen Moon S.S. Lam Eric S. K. Chin Spencer S.C. Tao , (2015),"Social media models, technologies, and applications", Industrial Management & Data Systems, Vol. 115 Iss 5 pp. 769 – 802

Fotini Patsioura Maro Vlachopoulou Vicky Manthou, (2009),"A new advertising effectiveness model for corporate advertising web sites", Benchmarking: An International Journal, Vol. 16 Iss 3 pp. 372 - 386

Frederic Crane, , G. (2010). «Marketing FO Entrepreneurs: ConceptandApplication for new .Venture. New York: SAGE Pablication, Inc

Frolova. Svetlana.(2014).THE ROLE OF ADVERTISING IN PROMOTING A PRODUCT, Thesis CENTRIA UNIVERSITY OF APPLIED SCIENCES ,Degree Programme in Industrial Management May 2014, p"1-61

Henry Boateng Abednego Feehi Okoe , (2015),"Consumers' attitude towards social media advertising and their behavioural response", Journal of Research in Interactive Marketing, Vol. 9 Iss 4 pp. 299 -312

Henry Boateng Abednego Feehi Okoe , (2015),"Consumers' attitude towards social media advertising and their behavioural response", Journal of Research in Interactive Marketing, Vol. 9 Iss 4 pp. 299 -312

Huang. Jinsong, Song Sub, , Liuning Zhouc, , , Xi Liu .(2013).Attitude Toward the Viral Ad: Expanding Traditional Advertising Models to Interactive Advertising, Journal of Interactive Marketing, Volume 27, Issue 1, February 2013, Pages 36–46

Idris, I., M.S.A. Yajid and A. Khatibi, 2009. Personal and Psychological Factors-Does it .Impact the Choice of Advertising Medium? Journal of Social Sciences, 5(2): 104-111

Ioannou Myria, Nikolaos Boukas, Evi Skoufari .(2014).Examining the role of advertising on the behaviour of co-operative bank consumers, Journal of Co-operative Organization and Management, Volume 2, Issue 1, June 2014, Pages 24-33

Jankowski. Jarosław, Przemysław Kazienko, Jarosław Wątróbski, Anna Lewandowska, Magdalena Zioło.(2016).Fuzzy multi-objective modeling of effectiveness and user experience in online advertising, Expert Systems with Applications, Volume 65, 15 December 2016, Pages 315-331

Kim, C., Park, S., Kwon, K., & Chang, W. (2012). An empirical test to measure the effectiveness of online advertising in online marketplaces using a hierarchical Bayes model. .Expert Systems with Applications (39), 117–128

Kim, ji-Hern., Hyun, yong j., (2010). " A model to investigate the influence of marketing – mix efforts and corporate image on brand equity in the IT software sector " . industrial . Marketing Management. No 40,2011, pp 424-438, Emerald Group Publishing Limited

Kim, S., & Jun, J. (2016). The impact of event advertising on attitudes and visit intentions. .Journal of Hospitality and Tourism Management, 29, 1-8

Kirby, J. and Marsden .(2006). Connected marketing : the viral, buzz and word of mouth .revolution, Oxford : Butterworth-Heinemann, 2006

Madera. Quetzali, Oscar Castillo ∗, Mario García-Valdez, Alejandra Mancilla..(2017). A method based on Interactive Evolutionary Computation and fuzzy logic for increasing the effectiveness of advertising campaigns, Information Sciences 414 (2017) 175–186

Marion Garaus and Udo Wagner , Anna-Maria B¨ack.(2017). The Effect of Media Multitasking on Advertising Message Effectiveness, Psychology & Marketing, Vol. 34(2):)138–156 (February 2017

Martín-Santana J.D., E. Reinares-Lara , P. Reinares-Lara .(2017).Influence of radio spokesperson gender and vocal pitch on advertising effectiveness: The role of listener gender, Spanish Journal of Marketing – ESIC, Volume 21, Issue 1, February 2017, Pages 63-71

McKay-Nesbitt, J., V. Manchanda, R., C. Smith, M., & A. Huhmann, B. (2011). Effects of age, need for cognition, and affective intensity on advertising effectiveness. Journal of Business Research (64), 12–17

Moch, Nils (2013). Competition in fragmented markets: New evidence from the German banking industry in the light of the subprime crisis. Journal of Banking &; Finance, Volume 37, Issue 8, Pp 2908–2919

Moorthy, S. and Hawkins, S. (2005). Advertising repetition and quality perception. Journal of .Business Research, 58, 354– 360

Mucan. Burcu, Özeltürkay. Eda Yaşa.(2014).Social Media Creates Competitive Advantages: How Turkish Banks Use This Power? A Content Analysis of Turkish Banks through Their Webpages, Procedia - Social and Behavioral Sciences, Volume 148, 25 August 2014, Pages 137-145

Muda, M.; Musa, R.; Niana-Mohamed, R. and Borhan, H. (2014). Celebrity Entrepreneur .Endorsement and Advertising Effectiveness. Social and Behavioral Sciences, 130, 11-20

Oladele. Kehinde Joseph, Ogunnaike, Akinbode Mosunmola, Aka Deborah .(2016). Effective Advertising: Tool For Achieving Client-Customer Relationships, Researchjournali's Journal of Media Studies Vol. 2 | No. 1 February | 2016

Olsen, G., Pracejus, J. W., & O'Guinn, T. (2012). Print advertising: White space. Journal of .Business Research (65), 855–860

Park, M. (2006). The compensatory effects of pictorial and verbal information for haptic information on consumer responses in non-store shopping environments. Unpublished Ph.D, .Ohio, Ohio state University

Park, T. (2008). Effective design of advertisement in handheld devices (Doctoral dissertation .)PURDUE UNIVERSITY

Park, T. (2008). Effectiveness Design of Advertisement in Handheld Devices. Dissertation for .doctoral degree, Purdue University, West Lafyette India

Pergelova, A., D. Prior and J. Rialp, 2010 Assessing Advertising Efficiency Does the Internet Play a Role? Journal of Advertising, 39(3): 39-54

Pilelienė Lina Viktorija Grigaliūnaitė , (2017)," The effect of female celebrity spokesperson in FMCG advertising: neuromarketing approach ", Journal of Consumer Marketing, Vol. 34 - .Iss 3 pp

Rosemary J. Averya, Matthew D. Eisenbergb, Kosali I. Simon .(2012).The impact of direct-to-consumer television and magazine advertising on antidepressant use, Journal of Health Economics, Volume 31, Issue 5, September 2012, Pages 705–718

Selviaridis. Kostas, Martin Spring b, Luis Araujo b .(2013). Provider involvement in business service definition: A typology, Industrial Marketing Management 42 (2013) 1398–1410

Strebinger .Andreas, Xiaoling Guo, Ferdinand Klauser, Peter Grant-Hayd .(2018).Is Multi-Ethnic Advertising a globally viable strategy for a Western luxury car brand? A mixed-method cross-cultural study, Journal of Business Research 82 (2018) 409–416

Suvittawat, A., (2015). "Commercial Bank Service Quality: Chonburi Province, Thailand". .International Journal of Management Sciences and Business Research. 4(2): 2226-8235

Tafani. Eric, Elyette Rouxa, Rainer Greifeneder .(2018).In the mood for action: When negative program-induced mood improves the behavioral effectiveness of TV commercials, Journal of Business Research 84 (2018) 125–140

Te'eni-Harari, T. (2014). "Clarifying the relationship between involvement variables and .advertising effectiveness among young people". Journal of Consum Policy. 37: 183–203

Thomas, S. and Kohli, S.C. (2011). Can Brand Image Move Upwards after Sideways? A .Strategic Approach to Brand Placements. Business Horizons, 54 (1): 41-49

Yahyaoui . Hamdi, Own .Hala S..(2018). Unsupervised clustering of service performance behaviors, Information Sciences 422 (2018) 558–571

Yuksel, A., Yuksel, F. & Bilim, Y. (2010) "Destination Attachment: Effects on Customer Satisfaction and Cognitive, Affective and Cognitive Loyalty", Journal Of Tourism Management, Vol.31, pp: 274-284.

STUDY OF NATURAL ANTIOXIDANT INHIBITORS FOR CORROSION PROTECTION OF MILD STEEL IN ACIDIC MEDIUM

Ms. Ibtihal Hassan AL-Bahrani, National University of Science and Technology, Sultanate of Oman.

ABSTRACT

Corrosion is a natural phenomena and it is a very big loss for structure as well as business; so corrosion minimization is one of the biggest areas of research. In this project; corrosion inhibition efficiency of berries juice extract was tested in mild steel sample against the sulfuric acid medium which was prepared from four types of berries; blackberries, blueberries, cranberries, and raspberries.

The experiment was designed by using Design Expert V 11 where immersed time, acid and berries juice extract concentrations were set as factors and a limited range was selected. The corrosion rate was set as a response. 20 runs were under taken and the experiments were done in the lab with maximum care.

The corrosion rate of 1020.374 mpy minimized to 597.659 mpy in presence of 5ml of extract and the efficiency of berries juice extract reached to 41.427%. The RSM gave 232.022 mpy corrosion rate as a minimum value and the optimum equation given by RSM was used for Genetic Algorithm analysis which gave 188.866 mpy corrosion rate at the optimizing conditions.

The surface and chemical characterization of the mild steel sample in presence and absence of berries juice extract were analyzed along with the relations between temperature, immersed time, acid and berries juice extract concentrations with response. The mixing tank was designed, calculations and drawing were presented.

KEYWORDS

Corrosion, Mild steel, Berries, Inhibitor, Weight loss, Sulfuric acid, RSM, SEM.

INTRODUCTION

Mild steel is used in most of our daily life applications as well as in industrial applications. In other hands, mild steel has low corrosion resistance which corrodes easily once it reacts with the surrounding environment; corrosion is a phenomenon where the mild steel reacts with the deteriorates and surrounding environment. Corrosion of mild steel is a continuous and constant problem where it severe financial and life loses as it leads to loss in the product, efficiency of equipment and failures in some cases (Roberge,2012).

The corrosion phenomenon controlled by adding a small amount of component to the corrosive environment which is a corrosion inhibitor. There are organic and inorganic inhibitors. The inorganic inhibitors are used in limit amount as they are toxic such as nitrites, molybate, nitrate, silicate, chromate, and phosphate (Cafferty,2010). Now a day, the corrosion engineers looking for cheap, renewable, friendly, readily available and ecologically acceptable inhibitors which are organic inhibitors which comes from the plants such as antioxidant components consisting of tannins, saponins, flavonoids, alkaloids and amino acids. Organic inhibitors consist of some heteroatoms such as oxygen, nitrogen, multi-bound, and sulphur atoms with the conjugated system which allows the absorption in the metal surface. Those atoms attract the absorption of some components on the metal surface and that done by the interaction of metals and by the electron of the conjugated system or by forming a bond with the free electron of heteroatom which forms a layer in the metal surface, as a result, it decreases the corrosion rate (Roberge,2012).

Berries are small pulpy fruits which have up to 9.2 mmol of antioxidant per 100g of berries. Berries can be used as a corrosion inhibitor as they are rich in antioxidant. The studies show that the berries juice extract of blueberries, cranberries, raspberries, and blackberries rich with antioxidant where the antioxidant capacity of them reach to 10 mg/ 100 g of blueberries (Michalska&Lysiak,2015), 58 mg/100g of cranberries (Blumberg,J.,et al,2013), 26.2mg/ 100 g of raspberries (Bobinaite,R.,Viskelis&Venskutonis,2015), and 21 mg/100 g of blackberries (Kaume,L.,et al,2012).

OBJECTIVES

To prepare anticorrosive extract (berries), to study corrosion in sample in presences and absences of the inhibitors, to determine the optimum conditions to inhibit corrosion and to design mixing tank.

METHADOLOGY, DESIGN AND EXPERIMENTATION

PREPARATION OF MILD STEEL SAMPLES

The mild steel sheets were cut into 2.5*3 cm^2 by using steel cutter and cleaned by surface cleaning machine to remove any corroded products in the metal surface then 4mm hole done by using drilling machine. Bunches for all metal samples (1-20) were made to differentiate between them. The drilled samples were washed by distilled water then rinsed with benzene and acetone respectively and dried with a hot air dryer. They were weighted using sensitive balance and kept in the desiccator until the time of usage.

PREPARATION OF CORROSIVE MEDIUM

The corrosive medium of different concentrations of H_2SO_4 prepared by using 36.8 N of the sulfuric acid.

PREPARATION OF BERRIES JUICE EXTRACT

137g, 209g, 282g, and 200g of raspberries, cranberries, blueberries, and blackberries respectively have been used to produce 309.5 ml of berries juice extract. The different types of berries washed by cold water, crashed in the electrical crasher and compressed mechanically by hand then filtered by cloth filter. The filtrate filtered again by using filter paper. The berries juice extract kept in the freezer at 0°C till the time of usage.

The efficiency of berries juice extract has been tested by running three experiments. The cleaned and weighted mild steel samples immersed in 0ml(blank), 2ml and 5ml of berries juice extract respectively with 50 ml of 1M H_2SO_4. After 2 hours; the samples have been taken out and cleaned by distilled water, benzene and acetone then dried by air dryer and weighted again. The corrosion rate and inhibitor concentration have been calculated by using equations (1) and (2) respectively.

$$Cr(mpy) = 22300\times\frac{\Delta W}{\rho\times A\times t} \quad \textbf{equation(1)}$$

$$\text{IE}(\%) = \frac{\text{initalcorrosionrate-finalcorrosionrate}}{\text{initalcorrosionrate}}\times 100 \quad \textbf{equation(2)}$$

DESIGN THE EXPERIMENT CONDITIONS

The experiments conditions have been designed by using Design Expert software-version 11 and following Response Surface Module(RSM); the range of the parameters have been given as inputs to the software which are inhibitor concentration (5-15 ml), immersed time (180-240 min) and acid concentration (1-2 M) and the response set as corrosion rate in mpy. Table(1) shows the experiment conditions given by the software.

Table(1): Designed runs given by RSM.

Run.No	Factor(A): inhibitor, ml	Factor(B): Time, min	Factor(C): Acid concentration, M
1	10	180	1
2	15	240	2
3	5	120	1
4	5	240	1
5	10	180	1.5
6	15	180	2
7	10	120	1.5
8	5	120	1.5
9	10	180	2
10	10	180	1.5
11	10	180	1.5
12	15	240	1
13	15	120	1
14	5	240	2
15	10	180	1.5
16	15	180	1.5
17	5	120	2
18	10	180	1.5
19	10	180	1.5
20	10	240	1.5

APPLYING THE DESIGNED RUNS IN THE LABORATORY

The weighted mild steel sample has been hanged by using a plastic thread in a 50 ml of sulfuric acid in presence of berries juice extract of known concentration and contact time as per the software; the batch has been kept in the shaker. After the contact time the mild steel sample cleaned by using 120 grade emery paper to remove any corroded products, washed by distilled water, rinsed by benzene and acetone respectively and dried by hot air dryer, weighted by using sensitive balance. The corrosion rate in mpy was calculated by using equation(1) where the weight losses results have been used.

STUDYING THE IMPACT OF TEMPERATURE IN CORROSION RATE

To study the impact of temperature in the corrosion rate; mild steel samples weighted and hanged for 2 hours in 50 ml of 1M sulfuric acid with 10ml of berries extract at different temperatures(30, 40, 50 and 60 °C) as shown in Table(2); each batch has been kept in the water bath shaker with heater to provide perfect contact between the solution and mild steel sample and to maintain the required temperature. Corrosion rate has been calculated by equation(1).

Table(2): Experiment conditions of temperature into corrosion rate.

SI.NO	Temperature (°C)	Inhibitor concentration (ml)	Acid concentration (M)	immersed time, (hours)
1	30	10	1	2
2	40	10	1	2
3	50	10	1	2
4	60	10	1	2

SURFACE AND CHEMICAL CHARACTERIZATION ANALYSIS

The mild steel surface of blank, absence and presence of berries juice extract samples have been studied by using Scanning Electron Microscopy (SEM) and Energy Dispersive Spectroscopy (EDS).

OPTIMIZATION STUDY BY GENETIC ALGORITHM

Matlab has been used for optimization where the genetic algorithm method has been followed. The actual equation form Design expert(RSM) has been used where "Optimtool" followed. The same limit of the conditions from RSM has been selected. The same previous procedure has been followed in the lab with maximum care by using those optimum conditions which given by the software, 1.001M of H_2SO_4 with 15 ml of berries juice extract immersed for 237.208 minutes.

DESIGN OF MIXING TANK

Design of mixing tank is a part of the project where the industrial scale mixing tank has been designed, the tank will be formed from high resistance material such as S.S to be able to resist the corrosive medium of sulfuric acid which will be mixed with berries juice extract. It is a container which used to mix various material (heterogeneous materials) together to form homogenous materials even the different viscosity materials can be mixed easily which provides continuous process, good control, easy cleaning, and simplest construction. It can be made from S.S material which provides a smooth surface and it can be easily cleaned (Petropedia,2018).

Parameters to be considered to design mixing tank are length, torque, weight, stiffness and hydraulic loading.(Dynamixianc,2015). The important factors in design mixing tank are tank type, tank volume, specific gravity, process and environment.(Ibryamoval,Y.et al,2009)

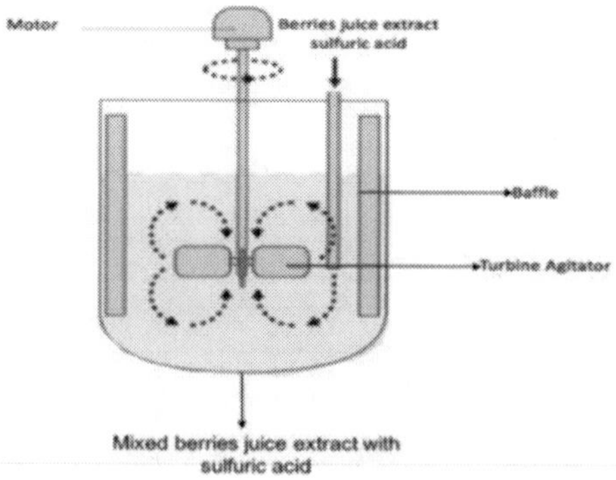

Figure(1): Mixing Tank(*Mixing Tank,*2018)

From the graph in Figure(2) Turbine agitator has been selected as it is suitable with the viscosity of the fluid and the tank volume. (Sinnot&Towler,2009)

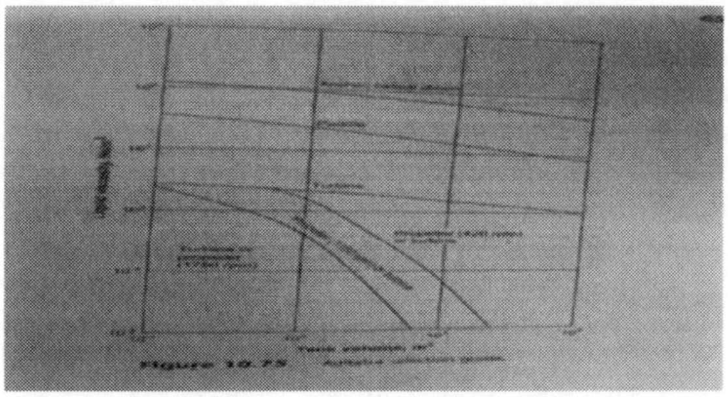

Figure(2): Graph of liquid viscosity Vs tank volume. (Sinnot&Towler,2009)

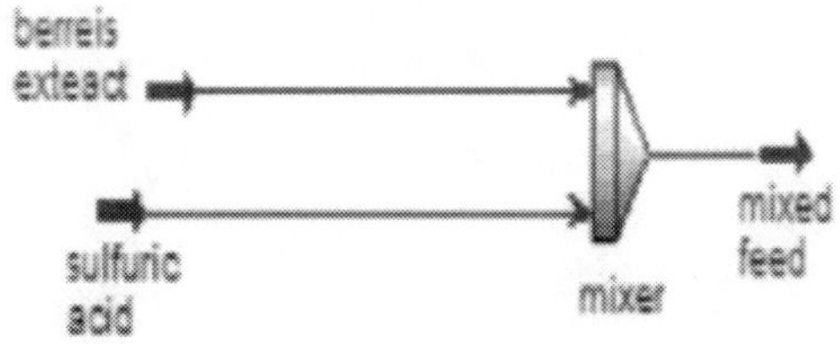

Figure(3): Result of Mixing Tank by Aspen Hysys.

Table(3): Calculated parameters for designing mixing tank

Tank diameter	m	3.5
Tank Height	m	1.81
Across sectional area	m^2	9.62
Volume of the tank	m^3	17.5
Impeller diameter	m	1.16
Length of paddle	m	0.29
Width of paddle	m	0.23
power	watt	204085
Reynolds Number	-	136486.6391

RESULTS AND ANALYSIS

EFFECT OF BERRIES JUICE EXTRACT CONCENTRATION INTO CORROSION RATE AND INHIBITOR EFFICIENCY

As shown in Table(4) the corrosion rate in the absence of berries juice extract inhibitor reach to the maximum (1020.374mpy); the corrosion rate has been minimized to 898.117mpy and 597.659mpy in presence of 2 and 5 ml of berries juice extract inhibitor respectively which means that berries juice extract inhibits the corrosion as the concentration of the berries juice extract increases. As well as the inhibitor efficiency reaches to 41.427% which obtained at 5ml of berries juice extract inhibitor in 1M of sulfuric acid immersed for 2 hours. The adsorption of the organic compounds such as ascorbic acid that present in the berries juice extract is blocking the active sites on the mild steel samples as it is forming a protective film on the metal surface as what kumar.A,2012 mention in his studies so the corrosion rate keeps decreasing as the amount of the berries juice extract is increasing. (Kumar.A,2012) Those compounds have heteroatoms such as oxygen which is a center of adsorption. In addition; It forms a coordinate bond as it has the ability to transfer the electrons to the vacant d-orbital of mild steel from the heteroatoms. (Ostovari, L. et al,2009)

Table(4): Experimental results of testing the efficiency of berries juice extract.

SI.NO	Inhibitor concentration (ml)	Corrosion rate (mpy)	Inhibitor efficiency (%)
1	0	1020.374	-
2	2	898.117	11.976
3	5	597.659	41.427

EFFECT OF IMMERSED TIME INTO CORROSION RATE AND INHIBITOR EFFICIENCY

The maximum corrosion rate reaches to 765.005mpy which obtained at 5ml of berries juice extract in the presence of 50 ml of 2 M of H_2SO_4 at 2 hours. Since the minimum corrosion rate of 232.022mpy obtained in presence of 5ml of berries juice extract with 50 ml of 1 M of H_2SO_4 immersed for 4 hours. The effect of immersed time into corrosion rate can be observed at low acid concentration and in the presence of berries juice extract, the corrosion rate is minimized at high immersed time. The decrease in the corrosion rate with the increase in immersed time is related to the adsorption of the corrosion product ($FeSO_4$) as shown in equation(3); According to Xianghong,L. et al ; the corrosion product of $FeSO_4$ becomes thicker with the immersed time, so it form a protective film (barrier) in the surface of mild steel (Xianghong, L. et al,2009).

$$Fe_{(s)}+H_2SO_{4(aq)} \rightarrow FeSO_{4(aq)}+H_{2(g)}$$, **equation(3)**

EFFECT OF ACID CONCENTRATION INTO CORROSION RATE

The effect of acid concentration into corrosion rate has been studied by using 1, 1.5 and 2 M of H_2SO_4 at different inhibitor concentrations and immersed time where the corrosion rate increase with the acid concentration at low inhibitor concentration and immersed time which reach to 765.005mpy at 2M H_2SO_4, 5ml of berries juice extract which immersed for 120 minutes as a maximum corrosion rate as shown in Table (3). In presence of the corrosive medium of H_2SO_4 with less amount of berries juice extract; the mild steel surface was exposed to H_2SO_4 medium where the corrosion reaction started as shown in equation(3) and $FeSO_4$ was formed with H_2 bubbles.

EFFECT OF TEMPERATURE INTO CORROSION RATE

From Table(5) can be noted that the rate of corrosion increases with temperature where it reaches to maximum 2407.994mpy at maximum temperature of 60°C. According to Ambrish, S. et al, at high temperature the berries juice extract molecules were partially absorbed in the mild steel surface so it was exposed to the corrosive medium of H_2SO_4, as a result, increasing the corrosion rate (Ambrish,S. et al,2012).

Table(5): Results of studying the effect of temperature into corrosion rate

SI.NO	Temperature,° C	Corrosion rate, mpy
1	30	701.507
2	40	1666.404
3	50	2008.861
4	60	2407.994

ANALYSIS BY RSM

Once the experimental results and corrosion rate calculated the RSM analysis has been done where the corrosion rate has been given to the software as a response. The response must be modified where the results fit with linear. The fit summary shows that the model F-value and P-value of 12.66 and 0.0001 respectively imply that the model is significant where A, B, C, and B^2 are the significant terms in this model only. 1.48 is a lack of fit P-value which indicates that the fit is not significant as there are pure errors.

0.5562 of the predicted R^2 is a reasonable agreement with 0.7105 of the adjusted R^2. The signal to noise ratio of 13.7252 of Adeq precision that the ratio grater than 4 is desirable. (Mycourses,2016)

The process variables are studied using ANOVA. The estimate coefficients in term of coded factors are A-Inhibitor, B-time, C-Acid and B^2. It is creating an accompanying condition for the sample with 1.0000 correlation coefficient. The coded and actual equations given by software are shown in equation(4) and (5) respectively.

- Corrosion rate(mpy)= 370.52-40.52(A)-126.97(B)+ 80.85(C) +66.58$(B)^2$ **equation(4)**
- Corrosion rate(mpy)= 1189.15487-8.10479(Inhibitor) -8.77421(Time) +161.69302 (Acid Conc)+0.018495 $(Time)^2$ **equation(5)**

The various relations can be studied from one factor graph; it can be noted that the corrosion rate increases with acid concentration and decreases with the increase in the amount of berries juice extract and immersed time as can be seen from the plots in Figure(3):Plot pf corrsoion rate Vs inhibitor concntration, Figure(4): plot of corrosion rate Vs acid concentration and Figure(5):Plot of corrosion rate Vs immersed time respectively.

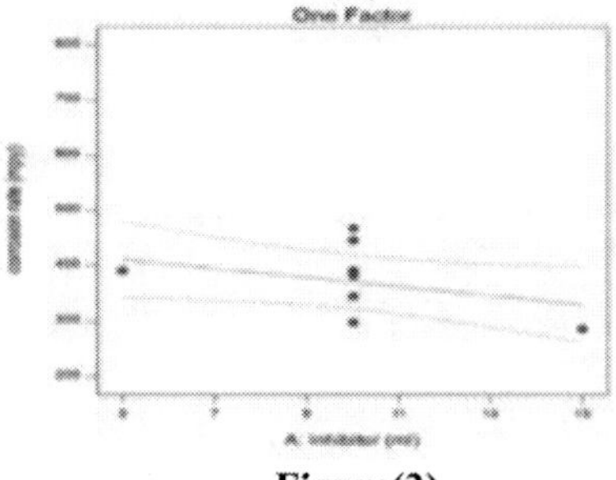

Figure(3)

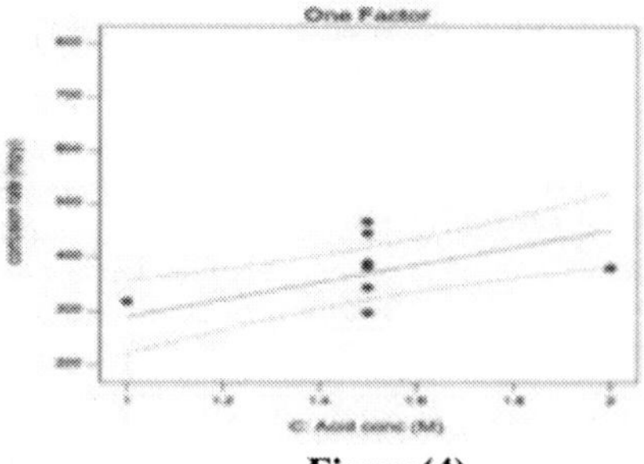

Figure(4)

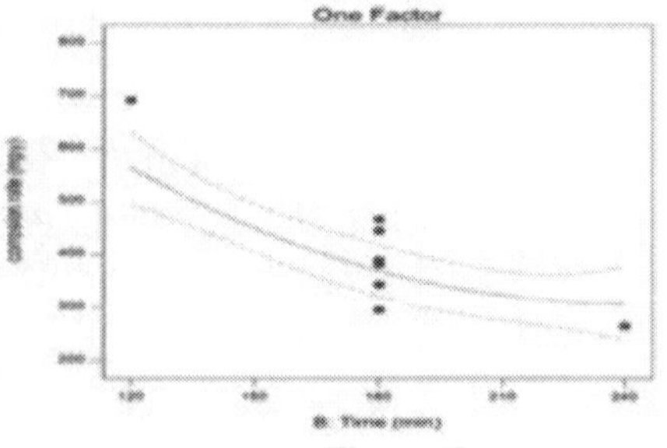

Figure(5)

The effects of two factors into the corrosion rate can be studied from the 3D plots as shown in Figure(6): 3D plot of corrosion rate Vs inhibitor concntration and acid concntration and Figure(7): 3D plot of corrosion rate Vs inhbitor concntration and immersed time.

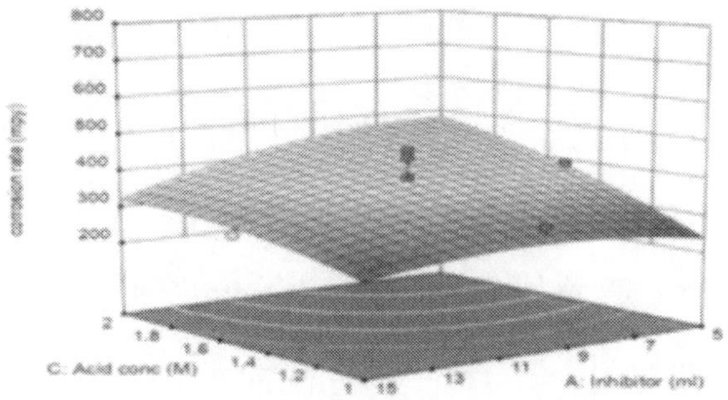

Figure(6)

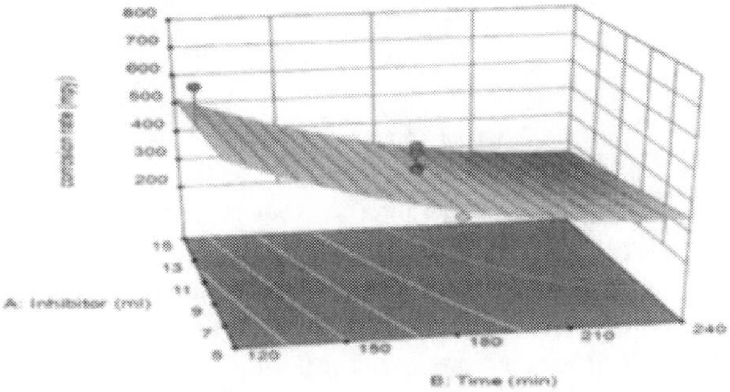

Figure(7)

The optimum conditions given by the RSM of 1M of H_2SO_4 with 5ml of berries juice extract inhibitor runs for 240 minutes gives 213.402mpy as a minimum corrosion rate but the experimental results at the same conditions gives 232.022mpy which means that they are too close as the suggested solution considered as an excellent optimum conditions.

SEM ANALYSIS

The SEM has been done for the blank mild steel sample for facilitating comparison as shown in Figure(8) which shows smooth mild steel surface compare to others. Rough surface in absence of berries juice extract is shown in Figure(9); where corrosion starting with the formation of H_2 bubbles as per the equation(3) which means that the surface of mild steel is exposed to the H_2SO_4 medium. In presence of 15 ml of berries juice extract, the surface of mild steel is protected by the adsorption of the organic compounds that are present in the berries juice extract where fewer cracks and pits can be noted and smooth surface is shown in Figure (10), which means that the mild steel surface is protected. (Raja, P. et al, 2008)

Figure(8)

Figure(9)

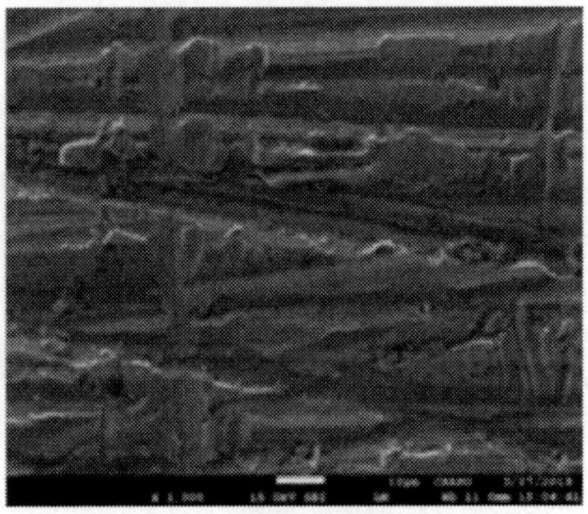

Figure(10)

EDS ANALYSIS

EDS has been used to study the chemical characterization or elemental analysis; Figure (11) shows the blank mild steel sample before being immersed in 1M H_2SO_4 in absence of berries juice extract where high Fe peak is shown, which indicates the presence of high amount of iron which reaches to 88 Wt% in the mild steel surface. Figure (12) shows the mild steel surface that has been immersed for 2 hours in 50 ml of 1M H_2SO_4 in absence of berries juice extract; the mild steel surface contains low amount of Fe (87.3 Wt%) compared to the blank mild steel sample. In Figure (13) the mild steel sample immersed for 2 hours in 50 ml of 1M H_2SO_4 in the presence of 5ml of berries juice extract, it shows high spark of Fe where the mild steel surface contains lower amount of Fe (84.8 Wt%) compared to Figure(11) and (12) which indicate lower iron dissolution rate in 1M H_2SO_4 medium containing berries juice extract (Alaneme, K. et al,2016). A high amount of oxygen and carbon are observed in presence of berries juice extract compared to the absence of berries juice extract. Si is shown in the absence and presence of berries juice extract which means that's it comes out to the surface during the reaction with sulfuric acid.

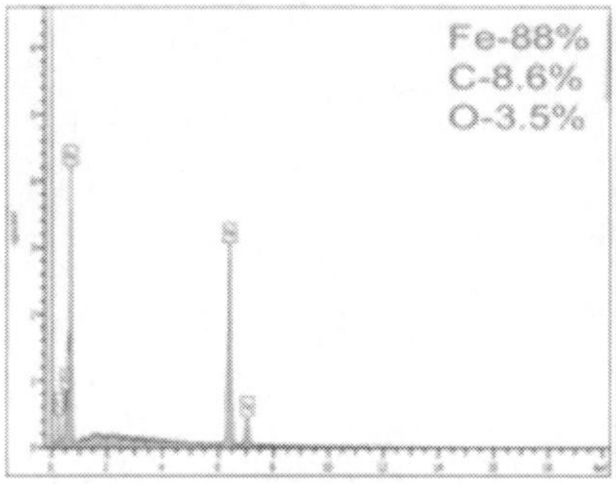

Figure(11)

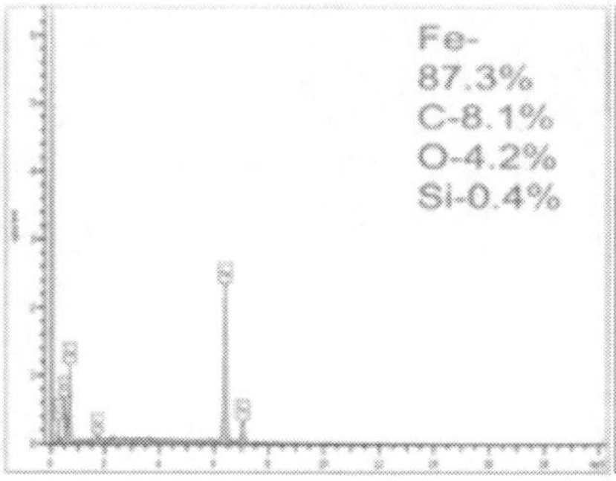

Figure(12)

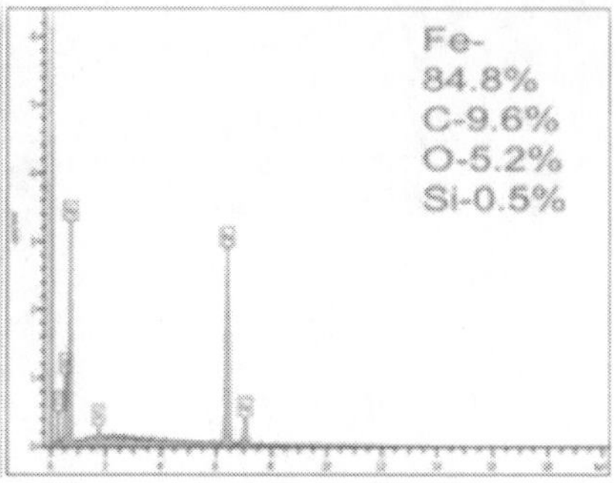

Figure(13)

GENETIC ALGORITHM OPTIMIZATION

The experimental result gives 288.367mpy as a corrosion rate. The software gives the minimum corrosion rate value of 188.866mpy from 50 runs as shown in Figure(14). A small difference can be noted as the genetic algorithm is not the exact fit with linear equations so the experimental and software results are too close to each other.

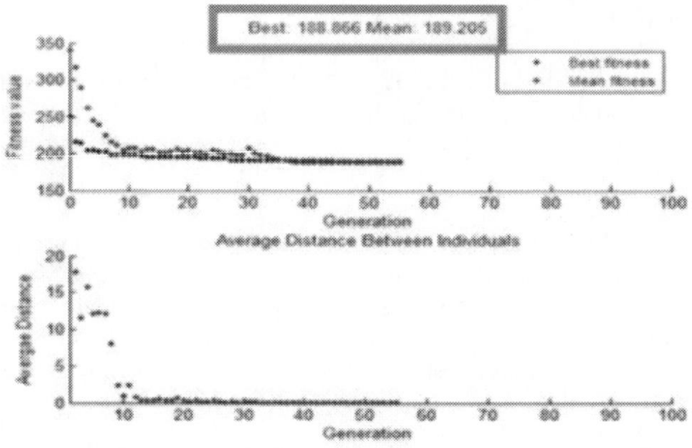

Figure(14): Genetic Algorithm optimization.

CONCLUSION

In this project, all the objectives have been achieved by designing and experimenting.
Firstly; the anticorrosive extract (berries) has been prepared, the prepared extract which contains 4 types of berries namely blueberries, blackberries, cranberries, and raspberries inhibit the mild steel corrosion in sulfuric acid medium with 41.427% efficiency in presence of 5 ml of berries extract.
Secondly, the mild steel corrosion in the presence and absence of berries juice extract were studied where the corrosion rate reaches to 1020.374mpy in absence of berries juice. By using 5 ml of berries juice extract the corrosion rate minimized to 232.022mpy.
Thirdly, the optimum conditions to inhibit the mild steel corrosion were studied by Design Expert V11. At the optimum conditions of 1M of H_2SO_4. 5ml of berries juice extract, 240 minutes and at room temperature the corrosion rate reaches to 232.022mpy as a minimum value.
The actual equation obtained from Design Expert V11 was later optimized using Genetic Algorithm. Genetic Algorithm shows that at 1.001M of H_2SO_4, 15 ml of berries juice extract and 237.208 minutes at room temperature, the corrosion rate is minimized to 188.866mpy.
Finally, mixing tank has been designed with 1.18m height and 2.5m diameter and a turbine agitator has been selected.

ACKNOWLEDGEMENT

I would like to thank god for giving me patience and power to work in my most important part of my life, my father, mother, my brother and my friends for their help and support. I would also like to express my sincere gratitude to Dr. Priy Brat Dwivedi for his support and help in developing my project idea, Mr. M.J.Varghese for his help in RSM analysis, Mr. Hatem Yazidi and Dr. Moammar Ali AlTubi for providing me the technical knowledge in Matlab optimization, Mr.Mohammed Salim Adeeb in providing me the technical help in the laboratory, Ganesh Nagraj Patil, Dr.Lakkim Setty

Rao, and Dr.Santosh Madhav Walke for their help in the calculations of design mixing tank. Last but not the least, I would like to thank National University of Science and Technology-College of Engineering for giving me this opportunity to complete the project in their laboratories and supporting me by providing me with the required chemicals.

REFERENCES

1.Alaneme,K. et al .,2016.Corrosion Inhibitory Properties of Elephant Grass(Pennisetum Purpureum) Extract Effect on Mild Steel Corrosion in 1M HCl Solution :*Alexandria Engineering Journal*.[e-journal]. 55 (8). p. 1069-1076. Available from: https://www.sciencedirect.com/science/article/pii/S1110016816300242.[Accessed: 8th May 2019].

2.Ambrish,S. et al .,2012. The Effect of Environmentally Bening Fruit Extract of Shahjan(Moringa Olifera) on the Corrosion of Mild Steel in Hydrochloric Acid Solution:*Journal of Chemical Engineering Communications*.[e-journal]. 199 (6). p. 63-77. Available from:https://www.tandfonline.com/doi/abs/10.1080/00986445.2011.570390.[Accessed: 6th May 2019].

3.Blumberg,J.,et al.,2013.Carnberries and Their Bioactive Constituents in Human Health:*Journal of Advances in Nutrition*.[e-journal]. 12 (7). p. 618-632. Available from: https://academic.oup.com/advances/article/4/6/618/4595566.[Accessed: 1th May 2019].

4.Bobinaite,R.,Viskelis,P.&Venskutonis,P.,2015.Chemical Composition of Raspberry Cultivars:*International Journal of Molecular Sciences*.[e-journal]. 16 (8). p. 18642-18663. Available from: https://www.researchgate.net/publication/306158828_Chemical_Composition_of_Raspberry_Rubus_spp_Cultivars.[Accessed: 1th May 2019].

5.Cafferty,E.,2010.*Introduction to Corrosion Science*.United States:Springer.
Roberge,P.,2012.*Handbook of Corrosion Engineering*.2nd edition.United States:McGraw-Hill company.
6.Dynamixianc,2015.*Optimal Mixer Design* .[Online]. Available from: https://www.dynamixinc.com/optimal-mixer-design. [Accessed: 14th December 2018]

7.Ibryamoval,Y. et al .,2009.Design of Mixers for Chemical and Biochemical Industry: *Trakia Journal of Sciences*.[e-journal]. 8 (6). p. 6-11. Available from: http://tru.uni-sz.bg/tsj/Vol8.Suppl.3.2010/Y.Ibryamova.pdf.[Accessed: 13th December 2018].

8.Kaume, L. et al.,2012. The Blackberry Fruit A Review on Its Composition and Chemistry, Metabolism and Bioavailability and Health Benefits:*Journal of Agricultural and Food Chemistry*.[e-journal]. 37 (7). p. 5716-5727. Available from: https://www.ncbi.nlm.nih.gov/pubmed/22082199.[Accessed: 6th May 2019].

9.Kumar.A,2012. Inhibition of Mild Steel Corrosion in HCl Solution Using Pipali(Piper Longum)Fruit Extract:*Arabian Journal for Science and Engineering*.[e-journal]. 38 (9). p. 85-97. Available from: https://link.springer.com/article/10.1007/s13369-012-0409-9.[Accessed: 9th May 2019].

10. Michalska,A.& Lysiak.G .,2015.Bioactive Compounds of Bluberries:*International Journal of Molecular Sciences*.[e-journal]. 16 (8). p. 18642-18663. Available from: https://www.ncbi.nlm.nih.gov/pmc/articles/PMC4581264/.[Accessed: 1th May 2019].

11.*Mixing Tank*,2018.[Online]. Available from: https://www.conceptdraw.com/examples/agitator-tank-symbol.[Accessed: 19th November 2018].

12. Mycourses,2016. *Response Surface Modelling-RSM*.[Online]. Available from:https://mycourses.aalto.fi/pluginfile.php/364340/mod_resource/content/1/BOS_Lecture3_2016.[Accessed: 4th May 2019].

13. Ostovari, L. et al.,2009.Corrosion Inhibition of Mild Steel in 1M HCl Soluition by henna extract:*Journal of Corrosion Science*.[e-journal]. 32 (9). p. 1935-1949. Available from: https://www.sciencedirect.com/science/article/pii/S0010938X09002157. [Accessed: 6th May 2019].

14. Petropedia,2018. *Mixing Tank*.[Online]. Available from: https://www.petropedia.com/definition/7742/mixing-tank.[Accessed: 11th December 2018].

15.Raja,P. et al.,2008.Inhibitive Effect of Black Pepper Extract on the Sulphuric Acid Corrosion of Mild Steel:*Journal of Materials Letters*.[e-journal]. 62 (9). p. 1977-2979. Available from: https://www.sciencedirect.com/science/article/abs/pii/S0167577X08001316.[Accessed: 8th May 2019].

16. Sinnot,R.&Towler,G.,2009.*Chemical Engineering Design*.5th edition.India:Elsevier.

17. Xianghong,L. et al.,2009.Inhibition Effect of 6-benzylaminopurine on the Corrosion of Cold Rolled Ssteel in H_2SO_4 Soluition:*Journal of Corrosion Science*.[e-journal]. 51 (9). p. 620-634. Available from: https://www.sciencedirect.com/science/article/pii/S0010938X09000080. [Accessed: 6th May 2019].

590434 Verification of Firefighters' Heuristics and Predictive Model Development through Big Data Analysis

Abstract:

The heuristics accumulated in the field activities of firefighters are reviewed through big data analysis of fire occurrences and emergency responses in Gyeonggi-do Province, Korea and expanded as predictive models to be utilized for proper fire prevention activities according to time, day, and target through quantitative modeling. Empirical rules with high sympathy are collected through direct interviews with firefighters. Among these, three types of high-priority rules are compared and analyzed for big data such as fire occurrence and damage history in Gyeonggi-do in 2018. Big data comparison analyses are conducted, including the number of fires and damages that occurred in Gyeonggi-do in 2018. Furthermore, fire occurrence patterns by region, day of the week, time of day, and building type are derived. In addition to using empirical rules that have been validated through research, relatively inexperienced firefighters also can make better decisions by relying on refined quantitative predictive models replacing empirical rules, including local government and time-based factors that reflect big fire occurrence data.

CONVERTING SHELTER-IN-PLACE LOCATIONS TO ZERO VULNERABILITY SAFE HAVENS

J. A. Rau
MineARC Systems America, LLC
4850 W. Ledbetter Drive
Dallas, Texas 75236
USA

A. Young
MineARC Systems, Pty Ltd
274 Welshpool Rd
Welshpool, WA 6106
Australia

M. McDermott
High Performance Building Solutions
7415 Folger Dr.
Charlotte, NC 28226
USA

CONVERTING SHELTER-IN-PLACE LOCATIONS TO ZERO VULNERABILITY SAFE HAVENS

ABSTRACT

While it is not uncommon to find shelter-in-place or safe haven locations designated across chemical facilities, what varies is the level of protection offered at these sites. Functionality can range from merely a muster point (shelter-in-place) to a positively pressurized room (safe haven). Either way, when determining the room's acceptable performance criteria, the industry generally applies ERPG-3[1] or a vulnerability factor. An acceptable vulnerability factor is often deemed the probability of a fatality being less than one percent.

In 2017, MineARC Systems and High Performance Building Solutions (HPBS) began working with a leading energy company to convert five shelter-in-place spaces to safe havens, which included additional life support systems that would ensure zero percent vulnerability. The objective of this paper is to examine the process to harden the space against contaminated air intrusion and what critical life support systems are essential for converting a shelter-in-place location to a safe haven.

INTRODUCTION

Emergency shelter-in-place (SIP) forms an integral part of any plant's broader emergency response plan. Explosions, fires, and the release of smoke or other toxic gases are some of the types of incidents that can occur despite high levels of planning and safety precautions[2].

Numerous government agencies and other organizations recommend shelter-in-place as a measure to reduce harm in the event of a chemical release. Any building may be used as a temporary measure to reduce health risk from exposure to the toxic materials by simply closing windows and doors and turning off ventilation fans and air conditioning/heating systems.

Three factors govern the effectiveness of a SIP location: 1) the tightness of the building or leakage rate, 2) the concentration of the toxic gas, and 3) the release duration. The tightness of the building is primarily based on building construction; however, temperature differential and wind speed also play a factor. Identification of the maximum toxic concentration occurs during a Facility Siting Study (FSS). FSSs are a hazard analysis that defines Maximum Credible Event (MCE) scenarios using a consequence-based approach to produce a quantitative risk assessment. Determining potential consequences is a necessary step in the process of developing a comprehensive safety plan. The FSS determines the expected release duration based on stored inventories of toxic materials. The higher the toxic concentration and longer the release duration, the less effective sheltering in place becomes.

While occupants may feel safe sheltering-in-place from the external toxic hazard, typical buildings have high leakage rates. A typical wood framed constructed building can have 3-5 air changes per hour (ACH50[3]), rendering it ineffective against most chemical releases. Additionally, sealing to an acceptable airtightness can be time consuming and expensive and even with professional air sealing it is very difficult to air seal existing structures to an acceptable tightness.

A case study for the conversion of five shelter-in-place spaces to zero percent vulnerability safe havens is presented. There are considerations for effectively sealing existing infrastructure and ensuring positive pressure to ensure contaminants do not infiltrate safe haven locations along with the critical element of life support. The critical elements of life support that are of primary importance to human survival in a safe haven are:

- Positive pressure
- Breathable air supply
- Supplemental O_2
- Removal of accumulated CO_2

---

[1] Emergency Response Planning Guidelines (ERPGs) are values developed by the AIHA Guideline Foundation's Emergency Response Planning Committee. ERPG-3 is the maximum airborne concentration below which it is believed nearly all individuals could be exposed for up to 1 hour without experiencing or developing life-threatening health effects.

2 The CSB is an independent federal agency charged with investigating industrial chemical accidents. Over the last 10 years they have completed investigations into 15 separate chemical release incidents.

[3] ACH50 is the number of times the air volume in a building changes per hour at 50 Pascal of pressure. Commercial air leakage rates are commonly reported at 75 Pascals, but this leakage study was conducted at 50 Pascals based on the targets included in the FSS report.

- Cooling and dehumidifying

CASE STUDY

In 2017 MineARC Systems and its project partner HPBS began working alongside a major energy company to convert five shelter-in-place locations into zero percent vulnerability safe havens.

The process started with the energy company completing its FSS, which included plume modeling to define the chemical release concentrations at each building location along with ideal shelter-in-place locations[4] and internal minimum positive pressure[5]. This information was used by MineARC to calculate indoor concentrations and occupant vulnerability to highlight that shelter-in-place alone would result in the vulnerability of greater than zero percent.

From there, HPBS conducted air infiltration testing following ASTM E779-10 to determine an air leakage baseline (see Table 6) for rooms that were designated to be sealed (pre-weatherization). A site assessment was then completed to review each location and determine its suitability for sealing and conversion to a safe haven. Firstly, the measured baseline infiltration rates were compared to the FEMA 453 estimates based on construction type[6] (see Figure 2). All surfaces of the room were examined for ease of conventional air sealing and mechanical equipment sealing. Rooms that included false floors, large amounts of electrical conduit, cabinetry, computer servers or windows etc. were deemed difficult to seal (i.e. control rooms). Secondly, the building or rooms open floor space was checked to ensure it could accommodate the specified number of people in an emergency. FEMA 453 section 1.6 recommends 20ft^2 per person for durations of less than 24 hours. With the FSS recommending 4 hours, it was deemed acceptable to have an unrestricted floor space of 6ft^2 per person.

Manual weatherization (air sealing) of the rooms included installing a cleanroom suspended ceiling system with gaskets and tiles that met the air permeance rate for an air barrier material, cleanroom airtight light fixtures with gaskets, sealing around electrical and plumbing, air sealing all visible air pathways to make sure the interior gypsum board was functioning as the air barrier, installing low-leakage HVAC dampers, installing an emergency shut off for the mechanical system and adding door weather-stripping and sweeps. Post-weatherization air leakage tests were conducted to determine the air infiltration rate of the SIP spaces after completion of the manual air sealing, cleanroom suspended ceiling system and HVAC work. The weatherization air sealing methods improved the air barrier integrity for all the proposed SIP spaces by between 57% and 65% (refer to Table 6).

Aerosol-based envelope sealing was used to reduce leakage rates even further. AeroBarrier® is a building envelope sealing technology that simultaneously measures and seals buildings. AeroBarrier® works by forcing an atomized sealant into orifices under pressure gradually sealing leaks. Rooms were sealed during night shift with personnel able to re-access the rooms by morning. The AeroBarrier® air sealing methods improved the air leakage for all the SIP spaces between 7% and 88% over the post manual weatherization leakage rates (refer to Table 7). AeroBarrier® is not effective for sealing large holes so manual air sealing is required for certain locations prior to using AeroBarrier®.

MineARC engineers once again calculated infiltration concentration rates along with the vulnerability rate. As the probability of a fatality was still higher than zero, MineARC then specified its AirBANK Pressure System and necessary carbon dioxide and oxygen control equipment.

SAFE HAVEN LIFE SUPPORT EQUIPMENT

Primary Life Supporting Atmosphere

A safe haven needs to provide a reliable breathable atmosphere independent of external toxic gas concentrations. Human breathing consumes oxygen (O_2) and expels carbon dioxide (CO_2), carbon monoxide (CO), water vapor, and heat. At complete rest, an average person exhales air at 17% O_2 and 3.2% CO_2. This breathing rate results in a 'resting' O_2 uptake of 0.3L/min of O_2 and a resting CO_2 discharge of 0.24L/min (Brake, Fellows, & Bates, 1999). Taking into consideration the above values, a person in a closed system such as a shelter-in-place succumbs to CO_2 poisoning well before O_2

4 Shelter-in-place location is beyond the scope of this paper, but should be determined based on the chemical hazard, facility layout, and location of personnel.

5 Positive pressure requirements are based on average wind speed at the facility. As wind speed increases so does the SIP leakage rate.

6 The FEMA 453 estimates are intended to estimate filter sizing when ASTM E779 air leakage testing is not conducted.

asphyxiation. Venter et al. (1998) found an average oxygen consumption for 12 entrapped individuals of 0.44 liters per minute, at STP, over 24 hours (including sleeping). A figure of 0.5 liters per minute over a non-sleeping entrapment period of 12 hours is a prudent, conservative design value. This corresponds to a metabolic rate of about 160W or 80W/m^2 for a typical person with a body skin area of about 2m^2, which is equivalent to a breathing rate of about 12.5 liters per minute of fresh air.

Consequently, life support systems are needed within a safe haven that can ensure positive pressure keeps contaminants outside, along with replacing the O_2 consumed, and removing exhaled CO_2, moisture, and heat to maintain the atmosphere within acceptable threshold limits.

Positive Pressure

Positive pressure is integral to prevent the ingress of airborne contaminants that may compromise the breathable atmosphere for occupants. Establishing a Class 1 Safe Room (FEMA 453) requires the sufficient release of filtered air from an external source, to form a positively pressurized zone.

Buildings are typically constructed to provide natural or forced ventilation to allow permanent occupancy and potentially combined with large exposed windows and surfaces which can present challenges for ensuring protection during a chemical release. When determining adequate positive pressure within a space, several structural factors need to be considered, including:

- Shelter volume
- HVAC (Heating, Ventilating, and Air Conditioning)
- Door and window tightness
- Existing opening such as vents and elevator shafts
- Indoor and outdoor temperature
- Wind speed and building orientation
- The external screening of the structure

Building design and multiple ingress points suggest the conversion of a smaller interior room within a building can provide a higher level of protection, than the whole building, as it is less subject to wind and buoyancy forces which induce the infiltration of external hazards.

Breathable air from compressed air cylinders venting directly into the SIP forms a positive internal pressure. Determining the amount of air required to reach the necessary level of positive pressure involves measuring the areas internal volume and the relationship between the internal and external atmospheric conditions.

After reaching required ACH (air changes per hour) or tightness, a positive-pressure management system must be sized for the expected shelter duration.

Breathable Air Supply

If the shelter-in-place is airtight, a supply of breathable air is necessary to support occupants over the required shelter duration. Unventilated SIPs that are tightly sealed cannot be occupied for long periods without the risk of high carbon dioxide levels. A secondary source of breathable air must be circulated within the shelter to meet occupancy over the required amount of time.

Clean, breathable air is supplied to a chemical, biological and radiological (CBR) safe room in a few ways: by trapping air inside the room and minimizing the air exchange, passing contaminated air through a filter (not practical for high concentration chemical releases or for chemicals that cannot be filtered) to purify it as it is supplied to the room, and the supplemental supply of oxygen through high pressure compressed air cylinders.

The air supply must contain enough oxygen in its mix, and not contain contaminants like carbon monoxide, carbon dioxide, water, or any other toxins that poison safe haven occupants during entrapment.

Table 1 lists international standards for flowrates for breathable compressed air for self-contained breathing apparatus (SCBA), supplied air respirators (SAR) and refuge chambers.

Table 1 – Standards for acceptable air supply requirements

Air supply requirements	Relevant Standard
Open-circuit positive-pressure SCBA, including all components and accessories except the air cylinder (shell) must resist the permeation and penetration of HD and GB chemical agents when tested for a period of 6 hours on an upper-torso manikin connected to a breathing machine operating at an airflow rate of 40 liters per minute (lpm), 36 respirations per minute, 1.1 liters tidal volume.	NIOSH 42 CFR Part 84 Respiratory Protective Devices
The flow rate should be at least 85L/min (3cfm) per person to remove expired CO2 and CO but at least 255L/min (9cfm) to provide adequate cooling and maintain comfortable conditions. The air inflow is assumed to be at a temperature of 26.5°C and 65% relative humidity	ITA Report No. 14 – Guideline for the provision of refuge chambers in tunnels under construction, May 2014
For manually operated blower type air-hose respirators, the blower shall be capable of delivering air to the space enclosed by the facepiece at a flow rate of at least 85 L/min.	AS/NZS 17:16:2003 Respiratory Protective Devices

The NIOSH standard of 40L/min is for SCBA apparatus (i.e. mouthpiece or face mask) where dilution of CO_2 is not a concern. Conversely, the ITA and Australian standards specify 85L/min and are concerned with both oxygen consumption and carbon dioxide dilution. It is important to note that three times the amount of air is required to maintain comfortable conditions inside a refuge chamber, when compared to simply breathing (Health & Safety Executive UK, 2006).

Unless the positive pressure system is supplying breathable air at a rate that maintains oxygen and compressed gas levels within acceptable ranges (dependent on safe haven occupancy and duration), a CO_2 scrubber and supplementary oxygen supply are required.

Supplemental O_2

O_2 is an essential requirement for cellular activity, and therefore, the absence of O_2 for relatively short periods results in death. Humans within an enclosed environment consume O_2 and if not replaced, would result in an irrespirable atmosphere. An O_2 concentration of less than 11% poses a high risk of death.

Retaining the oxygen levels between acceptable limits is critical. Various international standards outline acceptable O_2 limits for a breathable environment; examples are shown in Table 2.

Table 2 - Standards for allowable O_2 levels

Minimum O_2 Content (% by volume)	Maximum O_2 Content (% by volume)	Relevant Standard
18.5%	23.0%	Mines Safety and Health Administration 30 CFR Parts 7 and 75 Refuge Alternatives for Underground Coal Mines; Final Rule
19.5%	23.5%	Occupational Safety and Health Administration 29 CFR 1910.146 Permit-Required Confined Spaces for General Industry
19.5%	23.5%	Australian Standard AS 2865–2009 Safe working in a confined space

OSHA and Australian Standards for working in a confined space specify an acceptable lower limit of 19.5%, which is consistent with many other standards around the world. The slightly lower MSHA limit of 18.5% is based on a person at rest, not working, therefore, not likely to have trouble in breathing; conditions that are likely to be present in a refuge chamber.

O_2 itself is not flammable but supports and accelerates combustion. OSHA defines O_2 enriched mixtures or atmospheres as those containing more than 23.5% O_2 by volume (OSHA, 2013). In O_2-enriched atmospheres, the reactivity of O_2 significantly increases the risk of ignition and fire.

Within a safe haven, pure O_2 must be utilized in conjunction with a CO_2 removal system.

Removal of Accumulated CO_2

Protection within a safe haven is substantially increased by filtering CO_2 from the air. CO_2 is being produced continuously by the cells of our body and is a byproduct of respiration. The inability to expel this gas or an environment with high concentrations of CO_2 leads to CO_2 poisoning. Breathing air with more than 3% CO_2 can induce dizziness, headaches, and vision disturbance. Prolonged exposure at greater than 6% could result in unconsciousness and death. Listed in Table 3 are examples of international standards of satisfactory CO_2 removal requirements for a breathable environment.

Table 3 – Acceptable CO_2 removal requirements

CO_2 Removal Requirements	Relevant Standard
CO_2 removal components shall remove CO_2 at a rate of 1.08 ft^3 (30.5 L)/h/person.	Mines Safety and Health Administration 30 CFR Parts 7 and 75 Refuge Alternatives for Underground Coal Mines; Final Rule
OSHA has established a Permissible Exposure Limit (PEL) for CO2 of 5,000 parts per million (ppm) (0.5% CO2 in air) averaged over an 8-hour workday (time-weighted average or TWA.)	Occupational Safety and Health Administration
5.2.8 Atmospheric monitoring. Atmospheric monitoring There should be the capability to monitor temperature continuously, and for carbon monoxide, carbon dioxide, and oxygen in the chamber during use in both externally supported or standalone modes. The concentration of carbon dioxide should not exceed 10000 ppm.	ITA Report No. 14 – Guideline for the provision of refuge chambers in tunnels under construction, May 2014

The CO_2 that is exhaled by the occupants within the safe haven must be removed by, constant flushing by an external breathable air source (e.g. compressed air cylinders used for positive pressure) or by the regeneration of the internal atmosphere with a CO_2 absorbent (scrubber).

Different regulations around the world, state varied human production rates between 30–50L per hour per person. When considering an acceptable production rate; it must be noted that the rate of breathing is primarily triggered by the CO_2 content of inspired air with higher CO_2 levels triggering faster respiration rates (Brake, Fellows, & Bates, 1999) (i.e. if the maximum allowable CO_2 concentration inside the safe haven is 1%, a lower production rate is acceptable compared to a 3% CO_2 limit).

If CO_2 removal mechanisms are inadequate, the occupants can develop CO_2 toxicity by re-breathing their own exhaled CO_2.

Cooling and Dehumidifying

Thermal tolerability with a safe haven should be addressed during the design phase. Assumptions must be made in preparation for safe haven design in regard to the number of personnel, activity level, available hydration and heat production. The minimum airflow to maintain a comfortable environment is set by taking these assumptions into account as well as the input temperature and humidity of supplied air.

The healthy human body maintains its internal temperature around 37°C. An essential requirement for continued normal body function is that the deep body temperature is maintained within the narrow 1°C limit (Epstein, Moran, 2006). A change of body temperature exceeding 1°C occurs only during illness or when environmental conditions surpass the body's ability to cope with extreme temperatures.

When a person is subject to extreme heat and humidity over some time, the person may experience a very rapid progression of symptoms or exhibit a few of the less significant symptoms before falling victim to more severe forms of heat-related illness.

The acceptable cooling requirements for underground refuge chambers are defined in various international standards, examples of which are listed in the below table.

Table 4 – Standards for acceptable cooling requirements

Cooling Requirements	Relevant Standard
Simulated emergencies, in which a full complement of people has occupied a refuge chamber for a significant period, indicate that humidity and temperature can increase very rapidly to potentially heatstroke-inducing levels. Refrigerated air conditioning is strongly recommended for both externally supported and stand-alone refuge chambers to counter this potentially dangerous problem. Inevitably, this places a heavy demand on the stand-alone power supply, but there are systems available that can cope, and at an acceptable cost.	Department of Mines, Industry Regulations and Safety, 2005 Refuge Chambers in Underground metalliferous mines – guideline: Safety and Health Division, Department of Industry and Resources Western Australia, 32 pp
Apparent temperature in the fully occupied refuge chamber must not exceed 95 °F.	Mines Safety and Health Administration 30 CFR Parts 7 and 75 Refuge Alternatives for Underground Coal Mines; Final Rule
It should be possible to maintain an internal temperature not exceeding 30 °C wet bulb globe temperature for at least 24 hours	ITA Report No. 14 – Guideline for the provision of refuge chambers in tunnels under construction, May 2014

MSHA regulations specify a maximum internal apparent temperature[7] of 95°F (35°C). NIOSH specifies an equivalent maximum wet bulb temperature of 28.9°C (NIOSH, 2007).

The temperature maintained inside the safe haven is determined by the metabolic heat liberated by the occupants, dissipated to the external environment by the heat transfer characteristics of the room or building. The external temperature primarily influences the heat transfer while the humidity is the combination of moisture produced by the occupants (via exhaled air and sweat) and water liberated by CO_2 scrubbing chemicals.

It is important to remember that in most cases, before sheltering-in-place, the air conditioning system will need to be shut down (unless it is a split system). For short durations, artificial cooling is not a critical requirement unless external temperatures are extraordinarily high, and/or the safe haven's volume is small compared to its occupancy.

INFILTRATION TESTING

Background

High Performance Building Solutions Inc. (HPBS) was contracted to perform an air infiltration study for spaces identified as candidates to be used as shelter-in-place (SIP) in the event of an accidental chemical release. This SIP strategy uses the existing structure and its indoor atmosphere to separate individuals from a hazardous outdoor atmosphere temporarily.

A series of pre-weatherization air leakage tests were conducted to determine the baseline air infiltration rate of potential SIP spaces. Manual air sealing, a cleanroom suspended ceiling system and low-leakage HVAC dampers were installed to reduce the overall air infiltration of the SIP spaces and to close off air leakage pathways from non-protected spaces in the building through the HVAC system. Post-weatherization air leakage tests were conducted to determine the air infiltration rate of the SIP space after the completion of weatherization and HVAC work. The SIP spaces were additionally air sealed using the proprietary AeroBarrier® system to further reduce infiltration rates.

Scope

Air sealing work was completed, and low-leakage HVAC dampers were installed to reduce the overall air infiltration of the SIP spaces and to isolate the HVAC from non-protected spaces in the building. Post-weatherization air leakage tests were conducted to determine the air infiltration rate of the SIP space after the completion of air sealing, suspended ceiling and HVAC work. AeroBarrier® air sealing was conducted on SIP spaces as a pilot project on September 18 through September 21, 2017. Post AeroBarrier® air leakage tests were conducted on September 21 and September 22, 2017, to determine the air infiltration rate of the SIP space after the completion of the AeroBarrier® sealing.

7 Apparent temperature is the temperature equivalent perceived by humans, caused by the combined effects of air temperature and relative humidity.

Following ASTM E 779-10, pressurization air leakage tests were conducted on the following spaces recommended in the Facility Siting Study (FSS).

Table 5: SIP specifications

SIP Space	Occupancy	Floor Area (Square Feet)	Volume (Cubic Feet)
SIP Space 1	40	990	10,481
SIP Space 2	6	955	8,210
SIP Space 3	5	391	3,917
SIP Space 4	10	1,697	16,968
SIP Space 5	40	293	2,637

Air Leakage Test Results

Positive pressure air leakage test results are reported in four different ways:

CFM4 is the quantitative volume of air passing through an enclosure under a pressure difference of 4 Pascals (equivalent to a 5mph wind). This is a measurable physical quantity and the value recommended in the FSS to eliminate toxic gas ingress into the SIP spaces.

CFM25 is the quantitative volume of air passing through an enclosure under a pressure difference of 25 Pascals (0.1-in w.g.). FEMA 453 requires Class 1 Safe Havens be pressurized to 25 Pascals. This is a measurable physical property.

CFM50 is the quantitative volume of air passing through an enclosure under a pressure difference of 50 Pascals (0.2-in w.g.).

ACH50 is the quantitative volume of air passing through an enclosure under a pressure difference of 50 Pascals (0.2-in w.g.) normalized by the enclosure volume. ACH is the abbreviation for air changes per hour. It is the number of times the air volume in a building changes per hour at 50 Pascals of pressure.

Table 6 below shows the CFM50 and ACH50 for each of the SIP spaces pre-weatherization and post-weatherization along with the percentage reduction in leakage. SIP Space 1 and SIP Space 5 did not have pre-weatherization results as initially the whole building was tested. Smaller spaces were selected as the whole building was too large to feasibly convert to a SIP space (re-testing of the new SIP spaces was not completed). The control room, SIP space 4, was abandoned in favor of a smaller adjacent room due to high leakage post manual weatherization and the computers that could not be shut down, which prevented the use of AeroBarrier® in the space.

Table 6: Post-weatherization leakage results

Safe Haven Space	CFM50		ACH50		% Reduction
	Pre-Weatherization	Post-Weatherization	Pre-Weatherization	Post-Weatherization	
SIP Space 1	N/A	1,871	N/A	10.7	N/A
SIP Space 2	3,623	1,560	26	11.4	57%
SIP Space 3	385	135	6	2.1	65%
SIP Space 4	N/A	N/A	N/A	N/A	N/A
SIP Space 5	N/A	655	N/A	19.4	N/A

The AeroBarrier® air sealing was prescribed a target. According to the AeroBarrier® simultaneous air leakage calculations, all the spaces met and surpassed this airtightness target; AeroBarrier® sealing was halted at this point. Table 7 shows the CFM50 and ACH50 for each of the SIP spaces as tested by HPBS post AeroBarrier® along with the percentage reduction. The HPBS results were utilized to calculate the amount of positive pressure required to convert the rooms to positive pressure safe havens.

Table 7: Post AeroBarrier® leakage results

Safe Haven Space	CFM50		ACH50		% Reduction
	Post-Weatherization	Post AeroBarrier®	Post-Weatherization	Post AeroBarrier®	
SIP Space 1	1,871	455	10.7	2.6	76%
SIP Space 2	1,560	182	11.4	1.3	88%
SIP Space 3	135	125	2.1	1.9	7%
SIP Space 4	N/A	173	N/A	3.3	N/A
SIP Space 5	655	138	19.4	3.1	79%

The AeroBarrier® air sealing methods improved the air barrier integrity for all the proposed SIP spaces between 7% and 88% over the post manual weatherization leakage rates. Air leakage was still observed behind many electrical penetrations including HVAC thermostats, and fire and smoke alarms. These were covered with plastic to protect them from the AeroBarrier® Sealant but had hidden holes behind them.

Figure 1 displays the post AeroBarrier® leakage rate in comparison to the FEMA 453 estimates for air leakage rates based on construction types. FEMA 453 estimates are reported in square feet of floor area at 25 Pascals and do not take into consideration the ceiling height and volume of the space.

Figure 1: Post AeroBarrier® results compared to FEMA 453 estimates

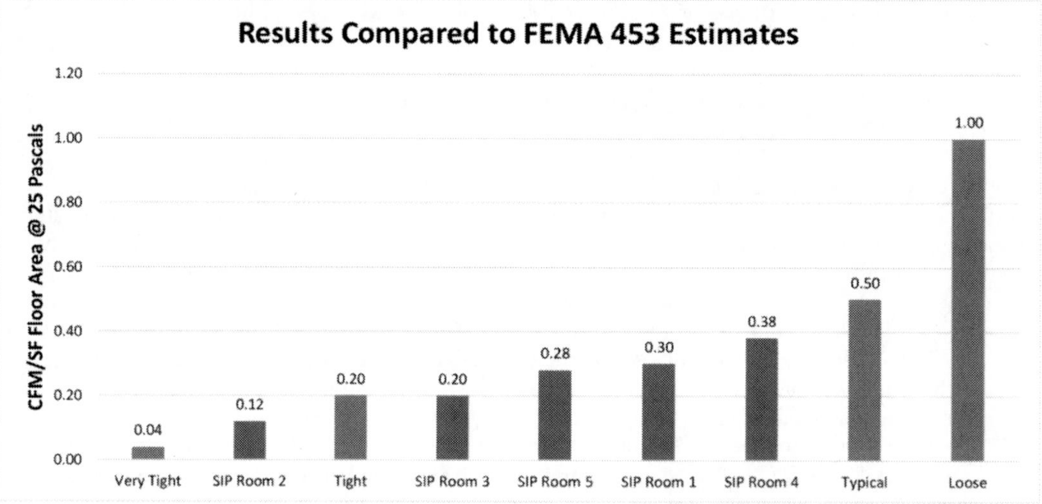

Figure 2 is an estimated leakage rate based on construction type to estimate the filtration and positive pressure system size under the FEMA 453 standard. All the SIP locations were between tight and typical except for SIP Room 2, which was a small prefabricated metal constructed portable building.

Figure 2: Estimated leakage rate per square foot of floor space to achieve 0.1 iwg (25Pa) overpressure

Construction Type	cfm per square foot of floor area
Very tight: 26-inch thick concrete walls and roof with no windows	0.04
Tight: 12-inch thick concrete or block walls and roof with tight windows and multiple, sealed penetrations	0.20
Typical: 12-inch thick concrete or block walls with gypsum wall board ceilings or composition roof and multiple, sealed penetrations	0.50
Loose: Wood-frame construction without special sealing measures	1.00

Table 8 shows the combined AeroBarrier®, and manual air sealing methods improved the airtightness for all the proposed SIP spaces between 68% and 95% over the original leakage rates.

Table 8: Air infiltration improvement for SIP spaces

Safe Haven Space	ACH50			% Reduction
	Pre-Weatherization	**Post-Weatherization**	**Post AeroBarrier®**	
SIP Space 1	N/A	10.7	2.6	76%
SIP Space 2	26.5	11.4	1.3	95%
SIP Space 3	5.9	2.1	1.9	68%
SIP Space 4	N/A	N/A	3.3	N/A
SIP Space 5	N/A	19.4	3.1	84%

Figure 3 compares the ACH50 pre-weatherization, post-weatherization and AeroBarrier® air infiltration results.

Figure 3: Pre-weatherization, post-weatherization and AeroBarrier® comparison

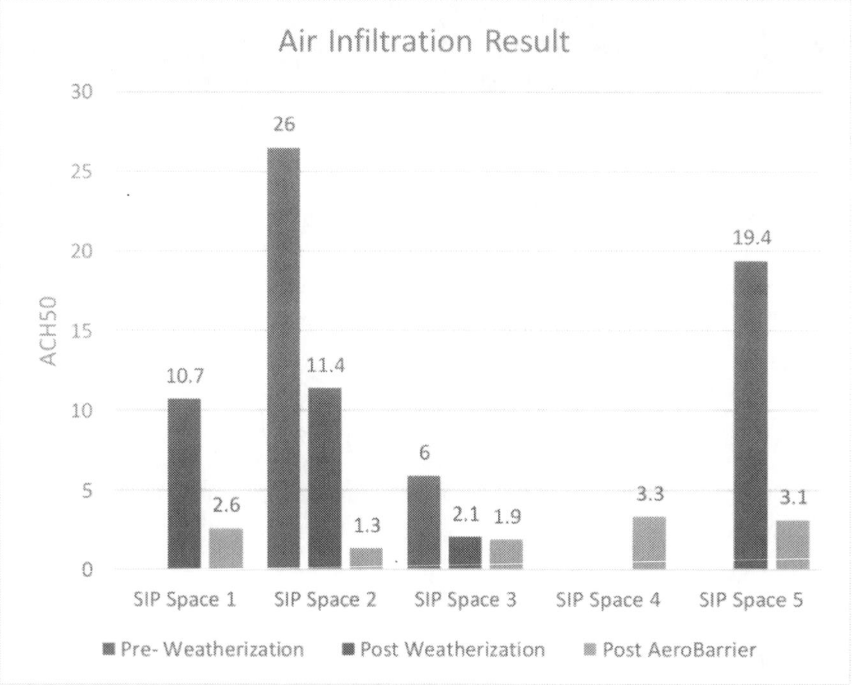

Life Support Systems

MineARC re-calculated the vulnerability for each SIP space based on the post AeroBarrier® leakage rates. All SIP spaces still had a vulnerability of 100% and therefore required the installation of independent life support systems to be deemed zero percent vulnerability safe havens.

Figure 4 shows the toxic gas concentration, duration of the release, and vulnerability for each SIP space. As the largest MCE was a full tank rupture the toxic concentrations were extremely high, but the duration was short at one minute.

Figure 4: SIP space vulnerability post AeroBarrier®

Safe Haven Location	TOXIC GAS				
	Worst Case Threat Scenario	Toxic External [ppm]	Duration of Reslease [min]	ACM	Vulnerability
SIP Space 1	Tank Rupture	585,000	1	0.032	100%
SIP Space 2	Tank Rupture	652,000	1	0.042	100%
SIP Space 3	Tank Rupture	660,000	1	0.007	100%
SIP Space 4	Tank Rupture	412,000	1	0.042	100%
SIP Space 5	Tank Rupture	600,000	1	0.028	100%

The leakage rate at 4 Pascals of pressure difference is slightly more than would occur under a 5 mile per hour wind used in the dispersion model in the FSS report based on climatic data for the facility (to maintain positive pressure and neutralize air infiltration). Post-weatherization infiltration rates in cubic feet per minute at 4 Pascals of pressure for each of the SIP spaces is shown in Table 9.

Table 9: Post AeroBarrier® air leakage results at CFM4

Safe Haven Space	Volume (Cubic Feet)	CFM4
SIP Space 1	10,481	99
SIP Space 2	8,210	32
SIP Space 3	3,917	24
SIP Space 4	3,110	45
SIP Space 5	2,637	21

The air leakage test results provide a mathematical formula that can be used to calculate the amount of breathable compressed air cylinders (437ft^3) that must be added to each candidate SIP space to maintain a 4 Pascal positive pressure to overcome a 5mph design wind coincident with a toxic release.

MineARC Systems was contracted to install life support systems to convert the SIP spaces to positive pressure safe havens. The MineARC AirBANK Positive Pressure System provides rapid pressurization, which is activated and maintained using the AirBANK Control via a simple human-machine interface (HMI) touch screen. MineARC's integrated Aura-FX Gas Monitor ensures breathable air automatically remains within acceptable limits. Additionally, the installation of supplementary oxygen and carbon dioxide scrubbing systems in safe havens, are as needed.

Figure 5 shows the number of 437 cubic feet cylinders required for each safe haven location. The minimum breathable air per person is 2.5cfm (70L/min) per person based on MineARC's experience and testing within the refuge chamber industry.

Figure 5: Quantity of cylinders required for 4Pa positive pressure

	SAFE HAVEN SPECIFICATIONS		POSITIVE PRESSURE (@ 4Pa)						
Safe Haven Location	Occupancy	Duration	CFM	CFM per Floor Space	ACH	No. Cylinders	No. Cylinder Racks	Breathable Air per Person [CFM]	Breathable Air per Person [LPM]
Safe Haven Space 1	40	4	99	0.10	0.6	55	5	2.5	70.8
Safe Haven Space 2	6	4	32	0.03	0.2	18	2	5.4	152.9
Safe Haven Space 3	5	4	24	0.06	0.4	14	2	4.8	135.9
Safe Haven Space 4	10	4	45	0.14	0.9	25	3	4.5	127.4
Safe Haven Space 5	40	4	21	0.07	0.5	12	2	0.6	17.0

The maximum CO_2 concentration is at 1% and O_2 between 19.5% and 23.5% based on OSHA limits (refer to Table 2). Due to the volume and number of people in Safe Haven Space 5, it required supplemental oxygen and carbon dioxide scrubbing to maintain the internal gas concentrations within specified limits. Three sodium chlorate oxygen candles and two AirGEN CO_2 scrubbers were included in the design.

Figure 6: Internal gas concentrations in safe havens

	SAFE HAVEN SPECIFICATIONS		POSITIVE PRESSURE (@ 4Pa)			O2 & CO2 CONCENTRATIONS (@ 4Pa)			
Safe Haven Location	Occupancy	Duration	CFM	CFM per Floor Space	ACH	Indoor O_2 %	No. O_2 Candles	Indoor CO_2 %	No. AirGEN Scrubbers
Safe Haven Space 1	40	4	99	0.10	0.6	20.45	0	0.50	0
Safe Haven Space 2	6	4	32	0.03	0.2	20.84	0	0.17	0
Safe Haven Space 3	5	4	24	0.06	0.4	20.77	0	0.23	0
Safe Haven Space 4	10	4	45	0.14	0.9	20.66	0	0.31	0
Safe Haven Space 5	40	4	21	0.07	0.5	18.67	3	2.04	2

In 2018 MineARC completed installation and training on the five safe haven spaces. Each safe haven was tested during the commissioning process to ensure it could maintain acceptable positive pressure.

CONCLUSIONS

The conversion of shelter-in-place (SIP) locations to zero percent vulnerability safe havens is a four-step process consisting of the following:

1) Identifying the correct SIP location based on ease of sealing and small volume based on the required occupancy.
2) Manual air sealing of SIP spaces.
3) AeroBarrier® sealing of SIP spaces.
4) Installation of life support equipment.

Identifying SIP spaces that can seal from the outset is crucial to a successful result. In the case study, three of the five original locations were abandoned in favor of rooms in adjacent locations due to high leakage rates or reducing the size to make the project more cost effective. Control rooms pose the most difficulty in trying to seal due to the false floor and a myriad of cable tray and pass-throughs installed for the computer systems. The smaller the room and the more robust the construction (i.e. prefabricated metal or concrete buildings), the better it seals.

Manual air sealing needs to be overseen by someone with familiarity in building tightness. While most of the tasks are general construction, standard contractors are not trained to work at the level of precision needed to ensure proper sealing. Even a tiny orifice or ceiling tile not sealed correctly can ruin a leakage test result. HPBS and MineARC Systems have developed a system for ensuring airtightness in construction methodology and by avoiding un-sealable spaces or spaces that are not cost effective to seal.

For future construction projects, incorporation of SIP spaces should occur during the design phase. Air barrier commissioning is standard in new construction and should include the following tasks at a minimum: design review of the plans with specifications for the air barrier and mechanical system - specifically as these relate to spaces designated for use as safe havens; quality assurance inspections during construction to ensure a continuous air barrier is in place; and final air leakage testing of the continuous air barrier to ensure the leakage rate of the safe haven spaces are within acceptable limits.

Complementary life support equipment is essential to sustain life within an airtight environment. For personnel to remain within the sealed safe haven, the environment must be habitable. Air is to be free of toxins such as carbon dioxide and acceptable levels of oxygen maintained. Positive pressure systems reduce the ingress of toxins and provide some or all of the required source of breathable air. Additional equipment such as compressed air, oxygen cylinders and air scrubbers are used to meet the shortfall. Furthermore, internal conditions must consider the human condition, supporting workable body temperature and humidity.

Reducing the risk of harm to personnel is a priority for emergency response planning. While the likelihood of an MCE occurring may be low, the risk to on-site personnel in the event of an incident is high. Converting shelter-in-place (SIP) locations to zero vulnerability safe havens improves emergency response.

REFERENCES

ASTM E779-10, *Standard Test Method for Determining Air Leakage Rate by Fan Pressurization*, ASTM International, West Conshohocken, PA, 2019, www.astm.org

Brake, Fellow, & Bates (1999). *Criteria for the design of emergency refuge stations for an underground metal mine*. Aus. IMM Proceedings, No.2, 1999, pp 1–7.

Department of Mines and Petroleum, 2013, *Refuge chambers in underground mines — guideline: Resources Safety*, Department of Mines and Petroleum, Western Australia, 41 pp.

Epstein, Y., & Moran D. (2006). *Thermal Comfort and the Heat Stress Indices,* Industrial Health, 44, 388–398.

FEMA 453. (2006). *Design Guidance for Shelters and Safe Rooms.* Federal Emergency Management Agency

Health & Safety Executive UK (2008). *Guidance and Information on the role and design of safe havens in arrangements for escape from mines.* Retrieved from http://www.hse.gov.uk/pubns/mines08.pdf

ITA Working Group 5 (2016). *Provisions for refuge chambers in tunnel projects under construction.* ITA Report N°014 - V2 / March 2018

Mines Safety and Health Administration (2008). *30 CFR Parts 7 and 75 Refuge Alternatives for Underground Coal Mines; Final Rule.*

National Institute for Occupational Safety and Health (2007). *Research Report on Refuge Alternatives for Underground Coal Mines.* Retrieved from http://www.cdc.gov/niosh/docket/archive/pdfs/NIOSH-125/125 ResearchReportonRefugeAlternatives.pdf

Occupational Safety & Health Administration (2013). *Permit-required confined spaces.* Retrieved from http://www.osha.gov/pls/oshaweb/owadisp.show_document?p_table=STANDARDS&p_id=9797

Suming Du. (2014). *On the effectiveness of shelter-in-place as a measure to reduce harm from atmospheric releases. Journal of Emergency Management, Vol. 12, No, 3, May/June 2014.*

Venter. J. (1998). *Portable Refuge Chambers: Air or Tomb in Underground Escape Strategies.* School for Mechanical and Materials Engineering, Potchefstoom University, Republic of South Africa

Worksafe Australia National Standard (2005). *Australian Standard AS: NZS 1716:2003 Respiratory protective devices.*

Yantosik. G. (2006). *Shelter-in-Place Protective Action Guide.* The Chemical Stockpile Emergency Preparedness Program (CSEPP). U.S. Department of Energy

589008 How the Field of Human Factors Helps Solve the Riddle of Serious Injuries and Fatalities

Abstract:

Process and occupational safety are often treated as distinct endeavors, with the former tending to focus on facility system function and the latter on individual worker wellbeing. Though other differences certainly exist between the two, both endeavors are generally purposed to prevent incidents that can cost lives, health, time, and money, in addition to employee morale and public trust. The overarching thesis of this paper is that the field of Human Factors is uniquely positioned to effect positive change in both of these major safety categories.

The backdrop for our argument is the recent finding that while injury and illness incidence rates published by the Bureau of Labor statistics have decreased in recent years, the occurrences of serious injuries and fatalities have declined at a significantly slower rate. We discuss possible explanations for this finding but note that regardless of the preferred explanation, the trend should motivate scientists, engineers, and other safety professionals to revisit long-held accident causation theories holding that near misses – these are incidents that could have resulted in serious injuries and fatalities but did not – are precursors of future serious incidents. While this idea still has merit, the just-mentioned statistics warrant reconsideration of the relations between serious and less-serious incidents to explain why the rates of the two are not decreasing in lockstep. We suggest these data indicate additional (human) factors are involved.

To address this riddle and to support our general thesis, we (a) review core tenets of process safety, occupational safety, and the field of Human Factors, (b) point out that the human is always an element common to both process and occupational safety contexts, and (c) discuss how preventing or mitigating human error can benefit both. We offer specific examples from both safety contexts to show how Human Factors perspectives and techniques can help. Examples touch on standards, policies, and other administrative controls (e.g., procedures), human-machine interfaces of both workplace hardware (e.g., valves) and software (e.g., SCADA systems), environments (e.g., physical, regulatory, and cultural), and communications between personnel (e.g., elements of team resource management). We conclude that fundamental Human Factors considerations can help prevent and mitigate human performance gaps and therefore increase the rate of reduction in serious injuries and fatalities in both process and occupational safety.

589233 Initial Conditions Used in Atmospheric Dispersion Models:Issues with Commonly Used Assumptions for Gas Jets and Plumes

Abstract:

When simulating jets and plumes, all atmospheric dispersion models require initial conditions that appropriately parameterize the behavior of a gas release to the atmosphere. Often, process conditions are such that loss of primary containment of a gas will result in choked flow conditions. A great deal of experimental and theoretical consideration has been given to the flow from stagnation conditions to choked flow conditions, but less experimental work has centered on describing the flow through the depressurization process to atmospheric pressure where atmospheric dispersion models would be applicable. Traditional modeling of the depressurization process has taken the simplified approach of assuming a (depressurized) "top hat" profile (laterally constant) with application of the principles of conservation of mass, momentum, and energy. Such assumptions are known to result in predicted depressurized velocities that exceed the (sonic) velocity at the point of loss of containment, and because of the energy balance, the associated temperatures are significantly less than temperatures at the choked conditions. Physically, these effects could possibly self-cancel because as jet velocity slows, the decrease in kinetic energy of the jet will increase the jet temperature. However, no atmospheric dispersion model used for routine hazard assessment purposes accounts for kinetic energy in the energy balance. The traditional modeling approaches of the depressurization process are reconsidered based on analysis of available data, and an alternative approach to modeling the depressurization process is proposed. Example cases showing the important differences between the two approaches are discussed.

Fault Tree Uncertainty Analysis

Raymond "Randy" Freeman
S&PP Consulting
12303 Lake Shore Ridge
Houston, TX 77041
Cell: 713 408 0357
Rafree@yahoo.com

For Presentation at the 2020 Spring National AIChE Meeting
Houston, Texas

March 30, 2020

ABSTRACT

Fault tree analysis (FTA) is a widely used methodology in the process industries. FTA is used for the development of failure mechanisms, computation of failure frequencies and the determination of the probability of failure on demand of safety systems. Much of the data used in a FTA study are uncertain. For example, the failure rate of a pump is often not known with great precision. Likewise the failure rates of instrumentation are often known only within some defined limits. The common practice, used by analysts in the quantification of a fault tree, is to use the most likely or best guess as to the needed failure rate data. The use of best guess values as data inputs to the quantification of a fault tree creates uncertainty in the computed results.

This paper presents a general methodology for the determination of the impact of uncertainty on the results of a fault tree study. The general methodology is based on the mathematics of propagation of error and variance contribution analysis. For a fault tree that contains 10 event minimal cut sets, the methodology will compute the variance in the probability (or frequency) of the top event with a maximum error of 5%. For 5 event minimal cut sets the error in variance of the top event probability (or frequency) will be no more than 2%. An example is presented to illustrate the application of the fault tree uncertainty analysis methodology to a real world problem.

Keywords

Quantitative Risk Assessment, Fault Tree Analysis, Uncertainty Analysis

1. Background

Fault tree analysis is a widely used method for representation of failure mechanisms in chemical plants. The methodology is described in the CCPS CPQRA guideline book [Ref. 1] and is extensively discussed in Chapter 9 of Lees [Ref. 2]. Lacking in these discussions is a systematic method to quantify the uncertainty of the results of the fault tree calculations.

Fault tree analysis has also been used in the aircraft and aerospace industries, nuclear power plant review and is recognized by the OSHA Process Safety Management (PSM) standard [Ref. 3] as a method for process hazards analysis.

2. Review of Error Propagation and Variance Contribution Analysis (VCA) Methodology

The mean and variance of a function of random variables can be approximated using the method described by Haugen [Ref. 4] and applied by Freeman [Ref. 5, 6, 7]. Define an arbitrary function of a set of random variables, x_i, as:

Let

$$Y = F(x_i) \qquad \text{(Eq 1)}$$

The mean of Y can be estimated using the following approximation:

$$E(Y) = F[\,E(x_i)] \qquad \text{(Eq 2)}$$

Where:

E(Y) = expected value of random variable Y = mean of Y

$E(x_i)$ = expected value of random variable x_i = mean of x_i

The variance of Y can likewise be estimated as:

$$V(Y) = \sum_{i=1}^{n} \left[\frac{\partial Y}{\partial x_i}\right]^2 V(x_i) \qquad \text{(Eq 3)}$$

Where:

$V(Y)$ = variance of random variable Y as defined above in Equation 1

$V(x_i)$ = variance of random variable x_i as defined above in Equation 1

Note that the variance is simply the square of the standard deviation. Using the variance will simplify the mathematics that is described below. The contribution of each independent random variable to the overall variance in the function is:

$$V(Y \text{ from } x_i) = \left[\frac{\partial Y}{\partial x_i}\right]^2 V(x_i) \qquad \text{(Eq 4)}$$

The relative contribution of each term to the overall variance V(Y) is a measure of the importance in the uncertainty in the particular random variable, x_i. In effect, this is a sensitivity analysis combined with a uncertainty evaluation. The variance contribution combines the sensitivity in the answer to changes in the uncertain random variable, x_i, with a measure of the uncertainty in the random variable, x_i. The overall variance in Y is found by summing the sensitivity weighted variances from each random variable.

Numerical Estimate of the Sensitivity

Let us return to the fundamental definition of the derivative.

As before, set y = F(xi)

$$\frac{\partial y}{\partial x} = \lim_{\Delta xi \to 0}\left[\frac{F(xi+\Delta xi) - F(xi)}{\Delta xi}\right] \qquad \text{(Eq 5)}$$

We are interested in a numerical estimate of the derivative. Therefore, we will make the following approximation.

$$\frac{\partial y}{\partial x} \sim \left[\frac{F(xi+\Delta xi) - F(xi)}{\Delta xi}\right] \qquad \text{(Eq 6)}$$

Linear Functions

Let us examine the approximation for a simple linear function.

Let Y = AX + B (Eq 7)

What is the sensitivity of Y to a change in variable X? By simple calculus:

$$\frac{\partial y}{\partial x} = A \quad \text{(Eq 8)}$$

We can also find the sensitivity using the numerical approximation as:

$$\frac{\partial y}{\partial x} \sim \left[\frac{F(xi+ \Delta xi)- F(xi)}{\Delta xi}\right] \quad \text{(Eq 9)}$$

$$\frac{\partial y}{\partial x} \sim \left[\frac{(A(xi+ \Delta xi)+ B)-(A(xi)+ B)}{\Delta xi}\right] \quad \text{(Eq 10)}$$

$$\frac{\partial y}{\partial x} \sim \left[\frac{(A(xi+ \Delta xi))-(A(xi))}{\Delta xi}\right] \quad \text{(Eq 11)}$$

$$\frac{\partial y}{\partial x} \sim \left[\frac{((A\, xi+A\, \Delta xi))-(A(xi))}{\Delta xi}\right] \quad \text{(Eq 12)}$$

$$\frac{\partial y}{\partial x} \sim \left[\frac{A\, \Delta xi}{\Delta xi}\right] \quad \text{(Eq 13)}$$

$$\frac{\partial y}{\partial x} = A \quad \text{(Eq 14)}$$

Note that for a simple linear equation the numerical result for the sensitivity is exactly the same as the analytical expression. There is no restriction on the size of the perturbation (Δxi) used in the calculations. Previously, Freeman [Ref 7] has suggested using a 10% perturbation in xi to do the sensitivity calculations. However any size of perturbation will work and generate the correct sensitivity.

This is a general result. As long as the function can be expressed as a linear function of a number of variables, xi, the sensitivity can be directly calculated by a simple perturbation of the function. The resulting sensitivity will be exactly the same as would be obtained using analytical methods.

Non-Linear Expressions

In the quantification of a fault tree, the top event frequency is computed using failure rates of the component devices and time that measures the interval between system validations. In a minimal cut set of two or more basic events, the time variable may be the same. For example the test interval for a valve may be the same for a pressure transmitter. A minimal cut set involving both the valve and the pressure transmitter would create a 2rd order term in the test interval. If the test interval is uncertain, how do we compute the sensitivity of the top event frequency with respect to the test interval? We will now explore how to deal with this type of problem when completing the uncertainty analysis.

As before, let

$$y = F(xi) = Axi^{n} + B \quad \text{(Eq 15)}$$

What is the sensitivity of Y to a change in variable X? By simple calculus:

$$\frac{\partial y}{\partial x} = nAxi^{n-1} \quad \text{(Eq 16)}$$

Let us evaluate the numerical approximation for this case. If n=2, the function of interest is:

$y = F(xi) = Axi^2 + B$ (Eq 17)

From basic calculus, the exact analytical expression for the sensitivity is:

$$\frac{\partial y}{\partial x} = 2\ A(xi) \quad \text{(Eq 18)}$$

If xi = 1

$$\frac{\partial y}{\partial x} = 2\ A \quad \text{(Eq 19)}$$

Again the approximation of the derivative is:

$$\frac{\partial y}{\partial x} \sim \left[\frac{(A(xi+\Delta xi)^2+B)-(Axi+B)}{\Delta xi}\right] \quad \text{(Eq 20)}$$

$$\frac{\partial y}{\partial x} \sim \left[\frac{F(xi+\Delta xi)-F(xi)}{\Delta xi}\right] \quad \text{(Eq 21)}$$

Inserting the equation for F(xi) we obtain:

$$\frac{\partial y}{\partial x} \sim \left[\frac{(A(xi+\Delta xi)^2+B)-(Axi+B)}{\Delta xi}\right] \quad \text{(Eq 22)}$$

$$\frac{\partial y}{\partial x} \sim \left[\frac{(A\ (xi^2+2\Delta xi+\Delta xi^2)\ +B)-(Axi+B)}{\Delta xi}\right] \quad \text{(Eq 23)}$$

$$\frac{\partial y}{\partial x} \sim \left[\frac{A\ (xi^2+2\Delta xi+\Delta xi^2)\ -Axi}{\Delta xi}\right] \quad \text{(Eq 24)}$$

$$\frac{\partial y}{\partial x} \sim A\left[\frac{(xi^2+2\Delta xi+\Delta xi^2)\ -xi}{\Delta xi}\right] \quad \text{(Eq 25)}$$

Setting xi =1, and using a 10% perturbation as $\Delta xi = 0.1\ xi = 0.1$

$$\frac{\partial y}{\partial x} \sim A\left[\frac{(1^2+(2)(0.1)1+0.1^2\ )\ -1}{0.1\ (1)}\right] \quad \text{(Eq 26)}$$

$$\frac{\partial y}{\partial x} \sim A\left[\frac{(2)(0.1)+0.01}{0.1}\right] \quad \text{(Eq 27)}$$

$$\frac{\partial y}{\partial x} \sim A\left[\frac{0.21}{0.1}\right] \quad \text{(Eq 28)}$$

$$\frac{\partial y}{\partial x} = 2.1\ A \quad \text{(Eq 29)}$$

Note that the numerical estimate (Eq 29) is approximately 5% greater than the analytical value (Eq 19). As shown in Table 1, for higher order equations the error will increase as the size of the perturbation becomes more important. Table 2 shows the impact of reducing the perturbation size to 1%. For 10th order variables (say a minimal cut set with 10 basic events) the error is 5%. For almost all engineering evaluations, a maximum error of 5% should be adequate.

For cases where the test frequency, repair time, or mission time appears in a minimal cut set 3 or more times, I suggest that the following perturbation in time be used to compute the sensitivity of the top event:

$$\Delta time = 0.01\ time \qquad \text{(Eq 30)}$$

The sensitivity of the top event frequency to all uncertain variables can be computed using a perturbation of 1%.

$$\Delta xi = 0.01\ xi \qquad \text{(Eq 31)}$$

3. Fault Tree Analysis

Fault trees are often used to analyze potential failures of an engineered system. A fault tree is a failure logic diagram that shows the relationship of device or system failures leading to an accident. Fault tree analysis is covered in detail in the CCPS CPQRA book [Ref. 1]. A fault tree is constructed by starting with a top event or condition and asking the question:

What conditions lead to this outcome?

Each condition in the next layer down is then analyzed by asking the same question. What conditions lead to this outcome? The logic of how these intermediate conditions (termed intermediate events) are related is expressed using AND and OR logic gates. An AND gate output is true if and only if (iff) all of the inputs to the AND gate are all true simultaneously. An OR gate output is true iff one of the inputs to the OR gate is true.

Once the fault tree logic structure is created, numerical data can be used to calculate the frequency or probability of the top event of concern occurring. For fault trees where a basic event is found in multiple branches of the tree, such as power failure, the logic structure of the tree must be simplified to remove the impact of the repeated events on the top event of concern. This simplification is done using Boolean Algebra. Each resulting grouping of basic events can cause the top event to occur. These groupings of basic events are called minimal cut sets. As a simple example consider the Boolean Algebra representation of a fault tree :

$$T = BE11 + BE1 \bullet BE2 + BE1 \bullet BE3 + BE2 \bullet BE9 + BE3 \bullet BE7 \qquad \text{(Eq 32)}$$

Where:

T = Fault Tree Top Event

BE11 = Basic Event 11

BE1 = Basic Event 1

BE2 = Basic Event 2

BE3 = Basic Event 3

BE7 = Basic Event 7

BE9 = Basic Event 9

+ = Boolean Algebra Addition Symbol

• = Boolean Algebra Multiplication Symbol

For this example the top event T, the minimal cut sets are:

BE11

BE1 BE2

BE1 BE3

BE2 BE9

BE3 BE7

For example the existence of basic events BE1 and BE3 guarantees that the top event T will occur. Publically available standard computer software such as SAPHIRE [Ref. 8] may be used to find the minimal cut sets of a large fault tree.

The representation of the top event in equation 32 can now be used to compute the probability or frequency of occurrence of the top event. If the devices describe by the basic events are non-repairable and are tested after a period of time TI, the probability of top event, T, may be calculated as:

$$\text{Prob (T)} = \lambda_{11}\,TI/2 + (\lambda_1\,TI/2)(\lambda_2\,TI/2) + (\lambda_1\,TI/2)(\lambda_3\,TI/2) + (\lambda_2\,TI/2)(\lambda_9\,TI/2) + (\lambda_3\,TI/2)(\lambda_7\,TI/2) \qquad \text{(Eq 33)}$$

Where:

λ_{11} = failure rate of device 11 in basic event BE11, failures per unit time

λ_1 = failure rate of device 1 in basic event BE1, failures per unit time

λ_2 = failure rate of device 2 in basic event BE5, failures per unit time

λ_3 = failure rate of device 3 in basic event BE3, failures per unit time

λ_7 = failure rate of device 7 in basic event BE7, failures per unit time

λ_9 = failure rate of device 9 in basic event BE9, failures per unit time

TI = test interval, time

Equation 33 assumes that the device failure rates are small and the product λ TI < 0.1. The other major assumption is that the values of all of the failure rates and test intervals are known. In reality the failure rate data are not known exactly with only a range or probability distribution representing the state of knowledge.

4. Example Problem

The following example is taken from the ISA technical report on fault tree analysis [Ref. 9]. This example should not be considered as a recommendation by either ISA or AIChE. The example is presented to allow for the demonstration of the methods that can be used in a fault tree study to quantify the uncertainty in the resulting calculated failure frequency or failure probability.

Consider the interlock on the intermediate storage tank T101 shown on Piping and Instrument Diagram in Figure 1. The interlock block diagram is shown in Figure 2. The interlock is intended to prevent an abnormal condition in the tank which could lead to a uncontrolled release of the tank contents. The process hazards analysis team has recommended that the interlock be designed to safety integrity level 2 (SIL 2) with a target probability of failure on demand (PFD) of no greater than 1E-2 or a risk reduction factor (RRF) of 100. Does the proposed interlock shown on Figure 1 meet the SIL 2 target? What is the uncertainty in the predicted PFD of the interlock?

A fault tree for the failure of this interlock is shown in Figure 3. A minimal cut set analysis has been completed and the resulting minimal cut sets are presented in Table 3. We can now write the Boolean Algebra equation that represents this fault tree as:

$$T = PE + TS1 \cdot TS2 + LS1 \cdot LS2 + FT1 \cdot FT2 + FT2 \cdot FT3 + FT1 \cdot FT3 + BV1 \cdot BV2 + BV1 \cdot SOL1 + BV2 \cdot SOL2 + SOL1 \cdot SOL2 + PT1 \cdot PT2 \qquad \text{(Eq 34)}$$

Assuming that the interlock is non-repairable until tested at time TI and using lamda-time (λT) approximation of the failure rate, the Boolean representation of the fault tree may now be converted to a failure probability model as:

$$\begin{aligned} Prob(T) = {} & Prob(PE) + (\lambda_{TS1}\, TI/2)(\lambda_{TS2}\, TI/2) + (\lambda_{LS1}\, TI/2)(\lambda_{LS2}\, TI/2) + (\lambda_{FT1}\, TI/2)(\lambda_{FT2}\, TI/2) + \\ & (\lambda_{FT2}\, TI/2)(\lambda_{FT3} TI/2) + (\lambda_{FT1}\, TI/2)(\lambda_{FT3}\, TI/2) + (\lambda_{BV1}\, TI/2)(\lambda_{BV2}\, TI/2) + \\ & (\lambda_{BV1}\, TI/2)(\lambda_{SOL1} TI/2) + (\lambda_{BV2}\, TI/2)(\lambda_{SOL2}\, TI/2) + \\ & (\lambda_{SOL1}\, TI/2)(\lambda_{SOL2}\, TI/2) + (\lambda_{PT1}\, TI/2)(\lambda_{PT2}\, TI/2) \qquad \text{(Eq 35)} \end{aligned}$$

Where:

Prob(PE) = failure probability of electronic logic solver, assumed to be constant at 5E-3.

λ_{TS1} = failure rate of temperature switch 1, failures per unit time

λ_{TS2} = failure rate of temperature switch 2, failures per unit time

λ_{LS1} = failure rate of level switch 1, failures per unit time

λ_{LS2} = failure rate of level switch 2, failures per unit time

λ_{FT1} = failure rate of flow transmitter 1, failures per unit time

λ_{FT2} = failure rate of flow transmitter 2, failures per unit time

λ_{FT3} = failure rate of flow transmitter 3, failures per unit time

λ_{BV1} = failure rate of block valve 1, failures per unit time

λ_{BV2} = failure rate of block valve 2, failures per unit time

λ_{SOL1} = failure rate of solenoid valve 1, failures per unit time

λ_{SOL2} = failure rate of solenoid valve 2, failures per unit time

λ_{PT1} = failure rate of pressure transmitter 1, failures per unit time

λ_{PT2} = failure rate of pressure transmitter 2, failures per unit time

TI = test interval, time

The electronic logic solver PE is assumed to have a constant failure probability of 5E-3.

Failure rate data for the other devices are present in Table 4. The mode (most likely value) of the failure rate is set equal to the point value taken from the ISA example [Ref. 9]. The triangular probability distribution (Appendix A) was assumed to represent the failure rate data for the equipment items. The upper and lower limits are based on the failure rate data of Appendix 4 of Smith [Ref. 10].

Using the device failure rates represented by the Table 4 column labeled as the mode and the minimal cut sets presented in Table 3, the ISA technical report [Ref.9, page 35] computes an interlock PFD as 7.5E-3 (RRF of 133). This would satisfy the requirement for a SIL 2 interlock {PFD of 1E-2 or a RRF of 100}. Now what is the uncertainty in the interlock PFD?

The first thing to remember is that the PFD should be calculated based on the mean value of the failure rate lambda (λ), not the mode of the failure rate. For a data set, the mean is the best single point representation of the data set. Table 5 presents the PFD of each minimal cut set using the mean of the device failure rate lambda (λ). The total of the PFD becomes 1.47E-2 (RRF of 68). This would not satisfy a SIL 2 interlock requirement.

Now we will compute the variance of the predicted probability of top event, T. To compute the variance of the probability of the top event, T, we will need to compute the sensitivity of the top event to each of the variables.

Sensitivity Using Numerical Perturbation Calculation

We will use the ISA interlock example previously analyzed by Freeman (Ref. 7). Using public data sources Freeman found that the probability of failure on demand (PFD) was 1.49E-2. To use the perturbation method for the evaluation of the sensitivities requires evaluating the impact of small changes in the uncertain input parameters. To enable these calculations to be completed numbers using 8 significant digits were used. The calculations were completed using Microsoft Excel which uses double-precision floating-point arithmetic compliant with IEEE 754 specification. Excel nominally carries 15 significant figures in calculations. We will complete the calculations with Excel displaying 5 significant digits and round the result at the end of the calculations to 4 significant digits. We will illustrate the calculations by looking at the Flow Transmitter, FT1.

a. First we recalculate the PFD of the example and find it to be 1.4594E-02. This is the base number that we will use in the perturbation calculations.
b. We now make a small change in mean failure rate of FT1. The mean failure rate is 8E-6 failures/hour. A 1% perturbation is used or an increase of 0.08E-6 failures/hour.

c. The perturbed value of the failure rate of FT1 becomes
(8 + 0.08) * 1E-6 failures/hour or 8.08E-6 failures/hour.
d. The perturbed value of the failure rate of FT1 is now used to re-calculate the PFD of the interlock. The resulting re-calculation finds that the PFD is 1.4618E-02
e. The change in the PFD from the base number due to the perturbation of the failure rate of FT1 is now computed as:

delta PFD = 1.4618E-02 - 1.4594E-02 = 2.4556E-05

f. The sensitivity of the PFD to a change in the failure rate of FT1 is now computed by dividing the change in the PFD by the change FT1 failure rate as:

sensitivity = (2.4556E-05)/(0.08E-6 failures/hour) = 307.0 hours
Result is rounded to 4 significant digits.

We now note that in the paper by Freeman (Ref. 7) that equation 32 presents the sensitivity for FT1 as:

$$\frac{\partial Prob(T)}{\partial \lambda \mathrm{FT1}} = 307.0 \text{ hr} \qquad \text{(Eq 32 of Ref 7)}$$

The answers are exactly the same. The sensitivity calculations for all of the devices are presented in Table 6. Note that in every case the sensitivity found using the analytical expression and the numerical estimate are the same. This is a general result which will be true for all basic event equipment failure rates found in a fault tree minimal cut set.

Once found, the sensitivities can now be used to compute the variance contribution of each particular device to the overall variance of the top event PFD in the same manner as Freeman previously presented.

The sensitivities of the top event probability are summarized in Table 6. We may now compute the variance contribution of each uncertain variable using Equations 1 – 4 in the methodology section of this paper. The relative variance contribution is expressed as a percent of the total variance of the top event probability. The top event variance is found to be 8.24E-6 (failures per hour)^2 or equivalently the top event standard deviation is 2.78E-3 failures per hour.

In this example the variance (measure of uncertainty) is dominated by the flow transmitters. The flow transmitters account for almost 2/3 of the uncertainty in the top event probability. Failures of the flow transmitters appear in 3 of the minimal cut sets. Since the uncertainty in the transmitter failure rates is large, the flow transmitters should contribute a significant amount to the variance of the fault tree top event.

We can use the variance in the top event to define the likelihood of achieving SIL 2 performance of the interlock. The calculations are done using the normal probability distribution. The normal distribution is tabulated as the standard normal distribution using a normalization factor Z.

The standard normal factor, Z, [Ref. 11] is defined as:

$$Z = \left[\frac{xi - E(x)}{\sigma}\right] \qquad \text{(Eq 36)}$$

Where:

σ = standard deviation. Note that the variance of a random variable is the square of the standard deviation of the random variable.

$E(x)$ = Expected value of the random variable x_i

For the example, the expected value of the probability of the top event of concern was calculated as

$E[Prob(T)] = 1.47E\text{-}2$ (Eq 37)

From Table 4, the standard deviation in the probability of the top event is

$\sigma_T = 2.78E\text{-}3$ (Eq 38)

The target PFD for a SIL 2 interlock must not be worse than 1E-2 (RRF of 100).

We compute the normal distribution Z factor at the SIL 2 target value as

$Z = (0.01 - 0.0147)/2.78E\text{-}3$ (Eq 39)

$Z = -1.69065$ (Eq 40)

Tabulations of standard normal distribution are presented in most statistics books such as Meyer [Ref. 11]. For this value of Z, the corresponding probability that the interlock will achieve SIL 2 performance is 4.55 percent. We can also compute the risk reduction that the proposed interlock design will achieve.

What is the interlock probability of failure on demand (PFD) that we can be 95% certain that the interlock will provide? At the 95% level, the corresponding Z factor is 1.65 [Ref 11]. We now compute the corresponding PFD of the interlock that corresponds to this Z factor.

$$Z = \left[\frac{x95\% - E[Prob(T)]}{\sigma}\right]$$ (Eq 41)

In this case:

$x_{95\%}$ = the PFD that we can be confident that the interlock will achieve

$\sigma_T = 2.78E\text{-}3$ (Eq 42)

$E[Prob(T)] = 1.47E\text{-}2$ (Eq 43)

$Z = 1.65$ (Eq 44)

Rearranging equation 40 to find $x_{95\%}$ we obtain:

$x_{95\%} = E[Prob(T)] + Z\,\sigma_T$ (Eq 45)

$x_{95\%} = 1.47E\text{-}2 + (1.65)(2.78E\text{-}3) = 0.0193$ (Eq 46)

or a 95% certain risk reduction factor of

$RRF = 1/PFD = 1/0.0193 = 52$ (Eq 47)

By the same logic there is 5% chance that the risk reduction is a PFD of 0.01 or an RRF of 99. In summary, this interlock is likely to perform as a mid-range SIL 1 interlock, not the SIL 2 interlock desired by process hazards analysis team.

5. Methodology for Fault Tree Uncertainty Analysis

We can now generalize the methods used in the above example to a proposed method for the evaluation of the uncertainty in a fault tree. The following step-by-step procedure is modified from that that previously published by Freeman (Ref 7) to indicate the use of numerical methods to compute the sensitivity. Step 8 shown in *italics* indicates the modification..

1. Create the fault tree using standard methods outline in the CPQRA book [Ref 1].
2. Determine minimal cut sets by hand or using standard computer software such as SAPHIRE [Ref 12].
3. Define the needed failure rate data for each basic event.
4. Define those basic events that are considered to be uncertain.
5. For the basic events that are uncertain, define the probability distribution and associated parameters needed to numerically define the probability distribution. Appendix B of this paper presents the needed description of 4 commonly used probability distributions. In many cases the triangular distribution will be selected. The minimum, maximum and mode (most likely) parameters will be needed for the triangular distribution.
6. Compute the mean of each of the uncertain variable
7. Compute the probability or frequency of the fault tree top event of interest using the mean value for each uncertain variable.
8. *Compute the sensitivity of the top event probability or frequency using the numerical methods outlined in this paper.*
9. Compute the variance of each of the uncertain variables
10. Compute the variance contribution of for each of the uncertain variables to the top event probability or frequency using equation 6.
11. Compute the total variance of the top event of interest probability or frequency by summing all of the contributions determined in Step 10.
12. Compute the variance contribution percent of each uncertain variable by dividing the variable contribution (Step 10) by the total variance of the top event of interest (Step 11).
13. Define the level of risk the project is willing to take. What chance will the project management accept for potential failure of the interlock to achieve the desired risk reduction? In the above example, I have used a 5% risk of failure or a 95% certainty that the interlock will achieve a risk reduction factor of 52. In the example, there is a very low probability that the interlock will achieve a SIL 2 PFD of 0.01 or an RRF of 100.

Alternately you can report the 90% range for the top event of interest. For the example the 90% range starts at the 5% RRF of 99 and there is a 95% certainty that that an RRF of at least 52 will be achieved. The expected risk reduction for this interlock is an RRF of 68.

6. Conclusions

A general procedure has been presented to quantify the uncertainty in calculations of failure frequencies using the fault tree methodology. The uncertainty analysis procedure is based on the application of propagation of error and variance contribution analysis techniques to the minimal cut sets created during a fault tree study. The procedure is simple and can be incorporated into standard fault tree analysis programs such as SAPHIRE [Ref 8]. This procedure is based on perturbation calculation of the frequency of the top event from changes in the uncertain variables. Incorporation of a numerical perturbation method into standard fault tree software such as SAPHIRE would be easy to implement and would provide the means for evaluation of the uncertainty in the quantified fault tree results.

For the example interlock presented in this paper, the uncertainty in the device failure rates degrades the predicted interlock SIL from meeting SIL 2 requirements to becoming a mid range SIL 1 interlock. This is a general result. Uncertainty in design parameters will reduce the likelihood that the equipment will achieve desired results. The design engineer has two choices:

- Accept the conclusion that the interlock will not perform as SIL 2
- Re-design the interlock using more reliable components to achieve the SIL 2 target

The variance contribution analysis shown in Table 7 indicates that the major contributor to the uncertainty in the interlock performance is the Flow Transmitters. Flow transmitters certified to SIL-2 or SIL-3 performance with lower failure rates could improve the calculated PFD of the interlock.

7. References

1. Center for Chemical Process Safety , *Guidelines for Chemical Process Quantitative Risk Analysis – 2rd edition*, American Institute of Chemical Engineers, New York, 2000.
2. Frank P Lees, Loss Prevention in the Process Industries – 2rd Edition, Butterworth, ISBN 0-7506-1547-8, Oxford, UK, 1996
3. US Department of Labor, Occupational Safety and Health Administration, CFR 1910.119 - Process Safety Management of Highly Hazardous Chemicals, 1910.119(e)(2)(vi).
4. Edward B. Haugen, *Probabilistic Approaches to Design*, John Wiley, New York, 1968
5. Raymond. Freeman and Angela Summers, "Evaluation of Uncertainty in Safety Integrity Level Calculations," *Process Safety Progress,* Published online in Wiley Online Library (wileyonlinelibrary.com) DOI 10.1002/prs. 11805, 2016
6. Raymond. Freeman, "General Method for Uncertainty Evaluation of Safety Integrity Level Calculations – Part 2 Analytical Methods," *Process Safety Process*, Published online in Wiley Online Library (wileyonlinelibrary.com) DOI 10.1002/prs 111915, October 2017
7. Raymond. Freeman ,"Error Propagation and Uncertainty Analysis: Application to Fault Tree Analysis", Process Safety Progress, Published Wiley Online Library (wileyonlinelibrary.com/journal/prs) DOI: https://doi.org/10.1002/prs.12080, June 29, 2019

8. US Nuclear Regulatory Commission, "Systems Analysis Programs for Hands-on Integrated Reliability Evaluations (SAPHIRE) – Version 8," NUREG/CR-7039, Vol 3, June 2011.
9. ISA, "Safety Instrumented Functions (SIF) – Safety Integrity Level (SIL) Evaluation Techniques Part 3: Determining the SIL of a SIF via Fault Tree Analysis, ISA Technical Report ISA-TR-84.00.02-2002 – Part 3, Approved 17 June 2002.
10. David J. Smith, *Reliability Maintainability and Risk – 8th Edition*, ISBN 978-0-08-096902-2, Elsevier, London, 2011
11. Paul L. Meyer, Introductory Probability and Statistical Applications, John Wiley, New York, 1972

Table 1. Comparison of Analytical Sensitivity with Numerical Approximation with Perturbation of 10% in xi for the equation: $y = F(xi) = Axi^n + B$

Order of Equation n	Equation for Analytical Sensitivity $\frac{\partial y}{\partial x} = nAxi^{n-1}$	Sensitivity with xi = 1 and A=1	Numerical Sensitivity with $\Delta xi = 0.1\ xi$	Error %
1	$\frac{\partial y}{\partial x} = 1A(xi)^0$	1	1	0
2	$\frac{\partial y}{\partial x} = 2A(xi)^1$	2	2.1	5
3	$\frac{\partial y}{\partial x} = 3A(xi)^2$	3	3.31	-10%
4	$\frac{\partial y}{\partial x} = 4A(xi)^3$	4	4.64	-16%
5	$\frac{\partial y}{\partial x} = 5A(xi)^4$	5	6.11	-22%
6	$\frac{\partial y}{\partial x} = 6A(xi)^5$	6	7.72	-29%
7	$\frac{\partial y}{\partial x} = 7A(xi)^6$	7	9.49	-36%
8	$\frac{\partial y}{\partial x} = 8A(xi)^7$	8	11.44	-43%
9	$\frac{\partial y}{\partial x} = 9A(xi)^8$	9	13.58	-51%
10	$\frac{\partial y}{\partial x} = 10A(xi)^9$	10	15.94	-59%

Error = (Analytical- Numerical)/Analytical

Table 2. Comparison of Analytical Sensitivity with Numerical Approximation with Perturbation of 1% in xi for the equation: $y = F(xi) = Axi^n + B$

Order of Equation n	Equation for Analytical Sensitivity $\frac{\partial y}{\partial x} = nAxi^{n-1}$	Sensitivity with xi = 1 and A=1	Numerical Sensitivity with $\Delta xi = 0.1\,xi$	Error %
1	$\frac{\partial y}{\partial x} = 1A(xi)^0$	1	1	0
2	$\frac{\partial y}{\partial x} = 2A(xi)^1$	2	2.01	-1%
3	$\frac{\partial y}{\partial x} = 3A(xi)^2$	3	3.03	-1%
4	$\frac{\partial y}{\partial x} = 4A(xi)^3$	4	4.06	-2%
5	$\frac{\partial y}{\partial x} = 5A(xi)^4$	5	5.10	-2%
6	$\frac{\partial y}{\partial x} = 6A(xi)^5$	6	6.15	-3%
7	$\frac{\partial y}{\partial x} = 7A(xi)^6$	7	7.21	-3%
8	$\frac{\partial y}{\partial x} = 8A(xi)^7$	8	8.29	-4%
9	$\frac{\partial y}{\partial x} = 9A(xi)^8$	9	9.37	-4%
10	$\frac{\partial y}{\partial x} = 10A(xi)^9$	10	10.46	-5%

Error = (Analytical- Numerical)/Analytical

Table 3. Basic Events in Minimal Cut Sets For Example

Cut Set No	BE 1	BE 2
1	PE	-
2	TS1	TS2
3	LS1	LS2
4*	FT1	FT2
5*	FT2	FT3
6*	FT1	FT3
7*	BV1	BV2
8*	BV1	SOL1
9*	BV2	SOL2
10*	SOL1	SOL2
11	PT1	PT2

* Minimal cut Set with same basic event as another cut set

Table 4. Lamda (λ) Failure Rate Data With Uncertainty Limits

Device	Minimum Failures per 1E6 Hours	Mode Failures per 1E6 Hours	Maximum Failures per 1E6 Hours	Mean Failures per 1E6 Hours	Lamda Variance (Failures per 1E6 Hours)^2
Flow Transmitters FT1, FT2 and FT3	1.0	2.9	20.0	8.0	18.3
Pressure Transmitters PT1 and PT2	1.0	2.3	20.0	7.8	18.8
Temperature Switch TS1 and TS2	3.0	7.6	20.0	10.2	12.9
Level Switch LS1 and LS2	2.0	4.6	20.0	8.9	15.8
Block Valves BV1 and BV2	0.2	2.3	10.0	4.2	4.4
Solenoid Valves SOL1 and SOL2	1.0	2.3	8.0	3.8	2.3

Table 5. Probability of Minimal Cut Sets Based on Mean Failure Rates

Cut Set No	BE 1	BE 2	PFD based on mean λ
1	PE	-	5.00E-03
2	TS1	TS2	2.00E-03
3	LS1	LS2	1.52E-03
4	FT1	FT2	1.23E-03
5	FT2	FT3	1.23E-03
6	FT1	FT3	1.23E-03
7	BV1	BV2	3.38E-04
8	BV1	SOL1	3.38E-04
9	BV2	SOL2	3.38E-04
10	SOL1	SOL2	2.77E-04
11	PT1	PT2	1.17E-03

Table 6. Sensitivity of Example System PFD to Basic Event Failure Rates Based on a 1% Perturbation

Basic Event Label	Device Type	Mean Failures per 1E6 Hours	Failure Perturbation of 1% of Mean Failures per 1E6 Hours	Perturbed Failure Rate Failures per 1E6 Hours	Total System PFD Using Perturbed Failure Rate of Basic Event	Change in Total System PFD	Sensitivity of PFD to Change in Failure Rate Hours
FT1	Flow Transmitter	8	0.08	8.08	1.461848E-02	2.455603E-05	307.0
FT2	Flow Transmitter	8	0.08	8.08	1.461848E-02	2.455603E-05	307.0
FT3	Flow Transmitter	8	0.08	8.08	1.461848E-02	2.455603E-05	307.0
PT1	Pressure Transmitter	7.8	0.078	7.878	1.460560E-02	1.167179E-05	149.6
PT2	Pressure Transmitter	7.8	0.078	7.878	1.460560E-02	1.167179E-05	149.6
TS1	Temperature Switch	10.2	0.102	10.302	1.461389E-02	1.995945E-05	195.7
TS2	Temperature Switch	10.2	0.102	10.302	1.461389E-02	1.995945E-05	195.7
LS1	Level Switch	8.9	0.089	8.989	1.460912E-02	1.519596E-05	170.7
LS2	Level Switch	8.9	0.089	8.989	1.460912E-02	1.519596E-05	170.7
BV1	Block Valves	4.2	0.042	4.242	1.460037E-02	6.445958E-06	153.5
BV2	Block Valves	4.2	0.042	4.242	1.460037E-02	6.445958E-06	153.5
SOL1	Solenoid Valves	3.8	0.038	3.838	1.459976E-02	5.832058E-06	153.5
SOL2	Solenoid Valves	3.8	0.038	3.838	1.459976E-02	5.832058E-06	153.5

Table 7. Variance of Top Event Probability Due to Uncertain Device Failure Rates

Device No	Device Type	Label	Sensitivity of Top Event Probability to Device Failure Rate $Hour^{-1}$	Variance of Device Failure Rate $(Failures\ per\ 1E6\ Hour)^2$	Variance Contribution to Top Event Variance	Contribution to Top Event Variance %
1	Logic Solver	PE	-	0	0.00E+00	0%
2	Temperature Switch	TS1	195.7	12.9	4.94E-07	6%
3	Temperature Switch	TS2	195.7	12.9	4.94E-07	6%
4	Level Switch	LS1	170.7	15.8	4.60E-07	6%
5	Level Switch	LS2	170.7	15.8	4.60E-07	6%
6	Flow Transmitter	FT1	307	18.3	1.72E-06	21%
7	Flow Transmitter	FT2	307	18.3	1.72E-06	21%
8	Flow Transmitter	FT3	307	18.3	1.72E-06	21%
9	Block Valve	BV1	153.5	4.4	1.04E-07	1%
10	Block Valve	BV2	153.5	4.4	1.04E-07	1%
11	Pressure Transmitter	PT1	149.6	18.8	4.21E-07	5%
12	Pressure Transmitter	PT2	149.6	18.8	4.21E-07	5%
13	Solenoid valve	SOL1	153.5	2.3	5.42E-08	1%
14	Solenoid valve	SOL2	153.5	2.3	5.42E-08	1%

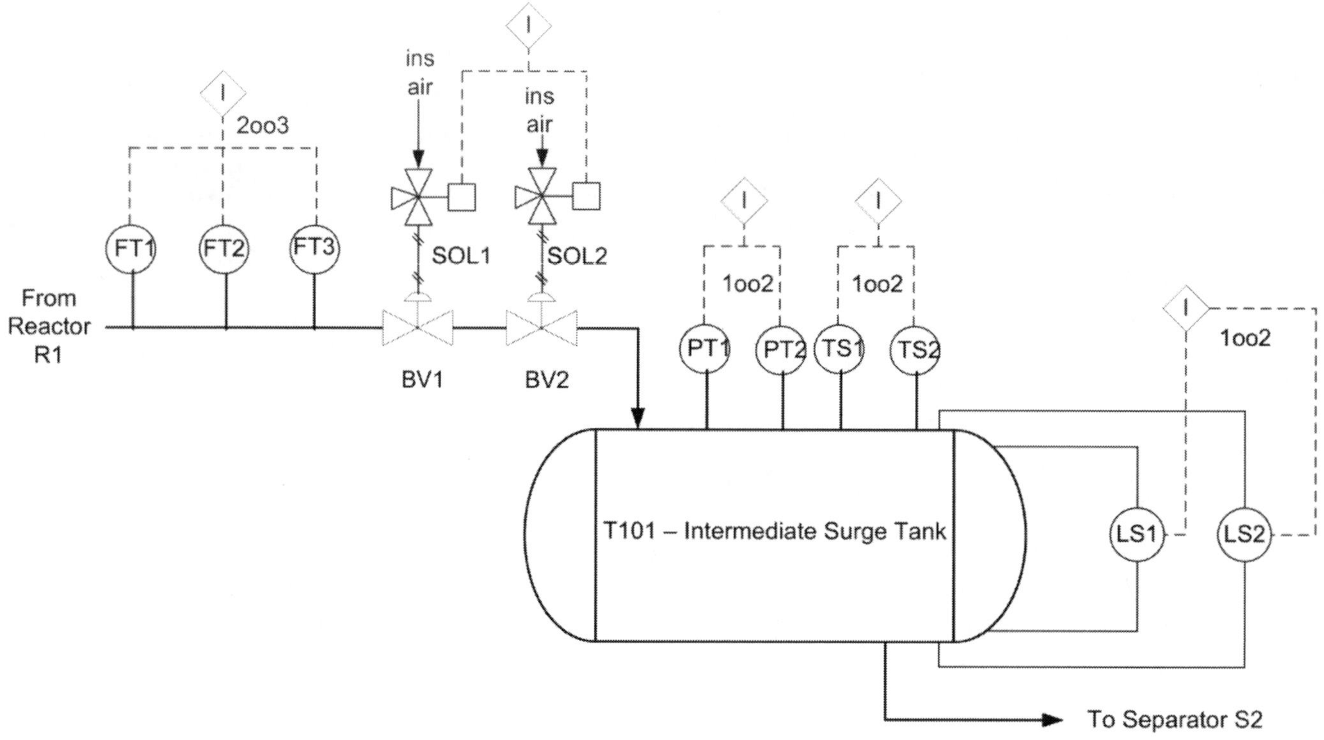

Figure 1. P&ID Diagram for Example

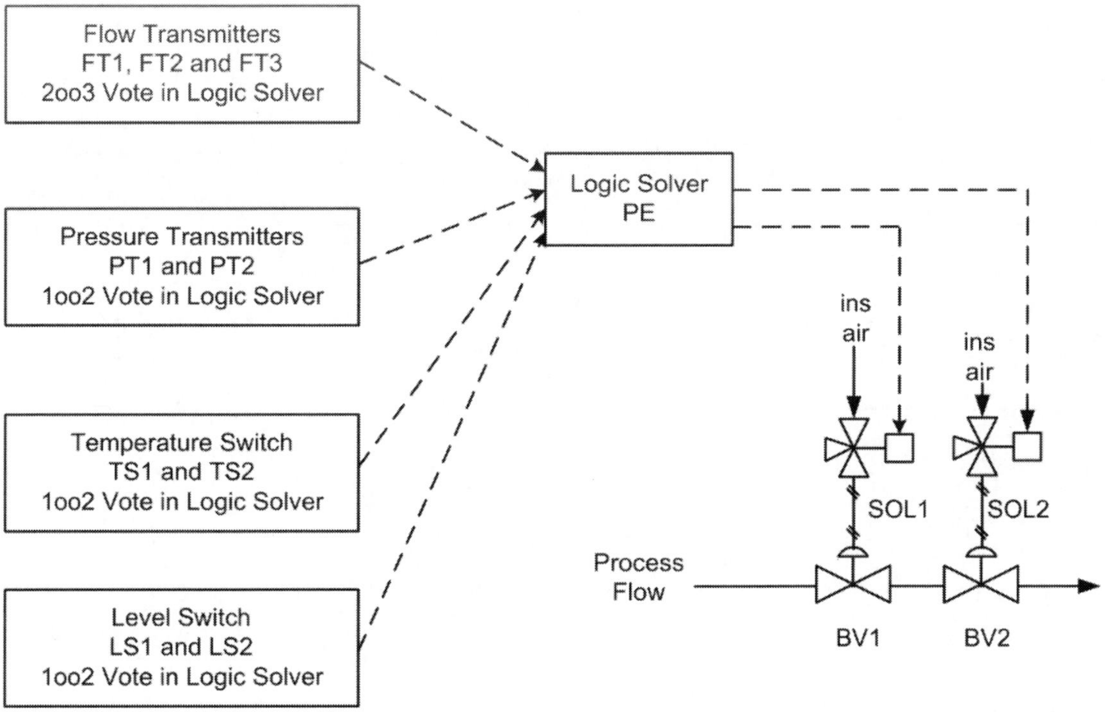

Figure 2. Interlock Block Diagram

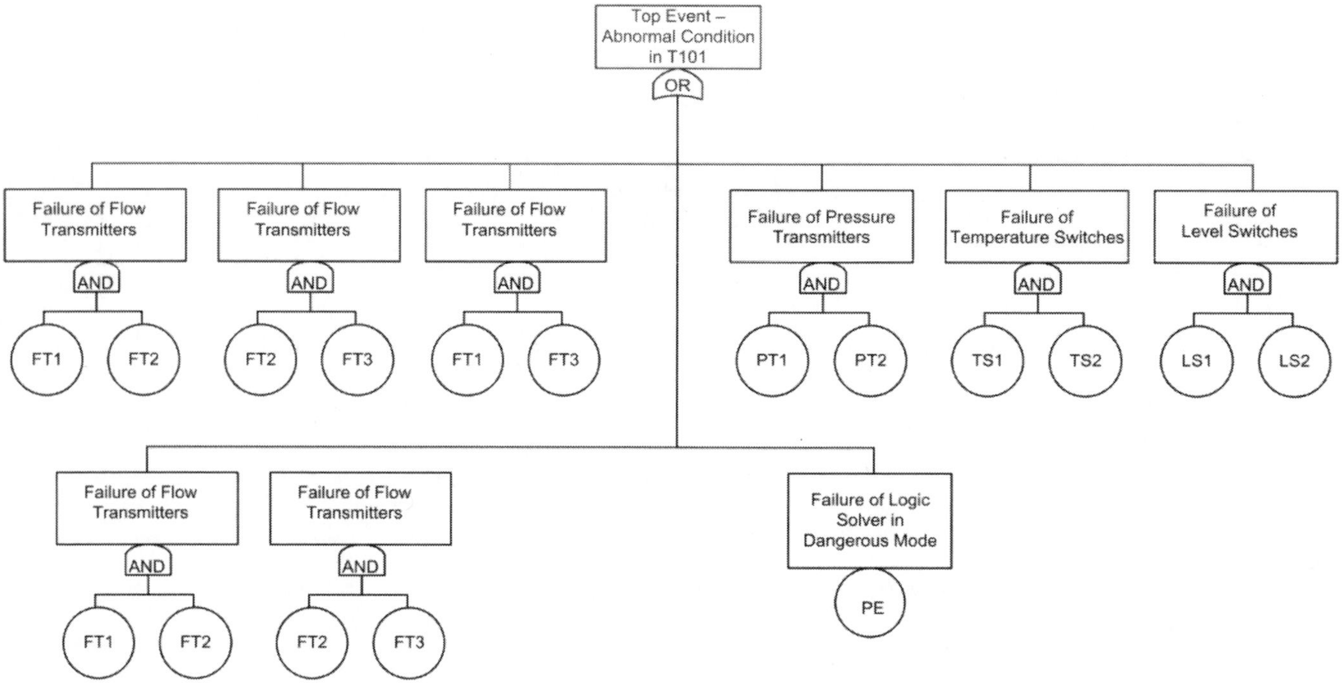

Figure 3. Fault Tree for Example Interlock

Appendix A – Commonly Encounter Probability Distributions

There are four probability distributions used to represent the failure rate of equipment items. These are:

1. Uniform Distribution
2. Triangular Distribution
3. Normal Distribution
4. Log-Normal Distribution

This appendix reviews the properties of these four probability distributions.

1. Uniform Distribution

The uniform distribution [Ref. A-1, pg 74; A-2, pg 687-688] is used when a variable is only known within a defined range. Figure A-1 presents a plot of the probability density function (pdf) for the uniform distribution. The distribution is defined using two parameters, A and B.

Where:

A = minimum value of the random variable

B = maximum value of the random variable

The mean or expected value of a random variable, x, that follows the uniform distribution is:

Mean = Expected Value – $E(x) = (A+B)/2$ (Eq A-1)

The variance of a random variable, x, that follows the uniform distribution is:

Variance = $V(x) = (B-A)^2/12$ (Eq A-2)

The standard deviation of a random variable, x, that follows the uniform distribution is:

Standard Deviation = $(\text{Variance})^{1/2} = (B-A)/(12)^{1/2}$ (Eq A-3)

2. Triangular Distribution

The triangular distribution [Ref. A-2, pg 686-687] is used when a variable is known to lie within a defined range and a "best guess" or estimate can be made as to the most likely value to the variable. Figure A-2 presents a plot of the probability density function (pdf) for the triangular distribution. The distribution is defined using three parameters, A. B and C.

Where:

A = minimum value of the random variable

B = maximum value of the random variable

C = most likely (mode) of the random variable

The mean or expected value of a random variable, x, that follows the triangular distribution is:

Mean = Expected Value – $E(x) = (A+B+C)/3$ (Eq A-4)

The variance of a random variable, x, that follows the triangular distribution is:

Variance = $V(x) = [A^2 + B^2 + C^2 - AB - AC - BC]/18$ (Eq A-5)

The standard deviation of a random variable, x, that follows the triangular distribution is:

Standard Deviation = $(\text{Variance})^{1/2} = \{[A^2 + B^2 + C^2 - AB - AC - BC]/18\}^{1/2}$ (Eq A-6)

3. Normal distribution

The normal distribution [Ref A-2, pg 665-666] is used when a variable when data analysis finds the normal distribution is the best model to describe the spread in the measured variable. Figure A-3 presents a plot of the probability density function (pdf) for the normal distribution. The distribution is defined using two parameters, μ and σ.

Where:

μ = mean of the measured data for variable

σ = standard deviation of the measured data for variable

The expected value of a random variable, x, that follows the normal distribution is:

Expected value = mean = $E(x) = \mu$ (Eq A-7)

The variance of a random variable, x, that follows the normal distribution is:

Variance = square of standard deviation = $V(x) = \sigma^2$ (Eq A-8)

The standard deviation of a random variable, x, that follows the normal distribution is:

Standard deviation = σ (Eq A-9)

4. Lognormal Distribution

The lognormal distribution [Ref. A- 2, pg 658-659] is used when a variable when data analysis finds the lognormal distribution is the best model to describe the spread in the measured variable. Figure A-4 presents a plot of the probability density function (pdf) for the lognormal distribution. Many variables in nature are lognormally distributed (for example the height of people). The log normal distribution is used when the variable of interest has a known physical lower limit of zero. A variable, x, is lognormally distributed if ln(x) is normally distributed. The lognormal distribution is defined using two parameters, μ_y and σ_y.

Let the original data be defined by the variable x with mean and standard deviation as:

μ_x = mean of the measured data for variable x

σ_x = standard deviation of the measured data for variable x

Define a new variable y as:

$y = Ln(x)$ (Eq A-10)

Then variable x is said to be lognormally distributed if y is normally distributed. We can now write the mean and standard deviation of y (μ_y and σ_y.) [Ref. A-3, A-4 and A-5] as

$$\mu_y = Ln\left[\mu_x^2 / ((\sigma_x^2 + \mu_x^2)^{\frac{1}{2}})\right] \qquad \text{(Eq A-11)}$$

$$\sigma_y = \left[Ln(\sigma_x^2 / \mu_x^2 + 1)\right]^{\frac{1}{2}} \qquad \text{(Eq A-12)}$$

Where:

Ln is the natural logarithm (base e) of the argument

Appendix A References:

A-1. P. L. Meyer, *Introductory Probability and Statistical Applications*, 2nd Edition, Addison-Wesley, Reading Mass, Library of Congress Catalog no. 75-104971, 1972, pp. 74

A-2. D. Vose, *Risk Analysis–A Quantitative Guide*, 3rd edition, Wiley, Chichester, West Sussex, England, ISBN 978-0-470-51284-5, 2008, pp. 686-687

A-3. "Lognormal mean and variance - MATLAB lognstat". www.mathworks.com. Retrieved 27 December 2018

A-4. M. Mood, F. A. Graybill, and D. C. Boes. *Introduction to the Theory of Statistics.* 3rd ed., New York: McGraw-Hill, ISBN 0-07-042864-6, 1974. pp. 117.

A-5. E.L. Crow and K. Shimizu, Lognormal Distributions – Theory and Applications, Marcel Dekker, New York, ISBN 0-8247-7803-0, 1988, pp. 9

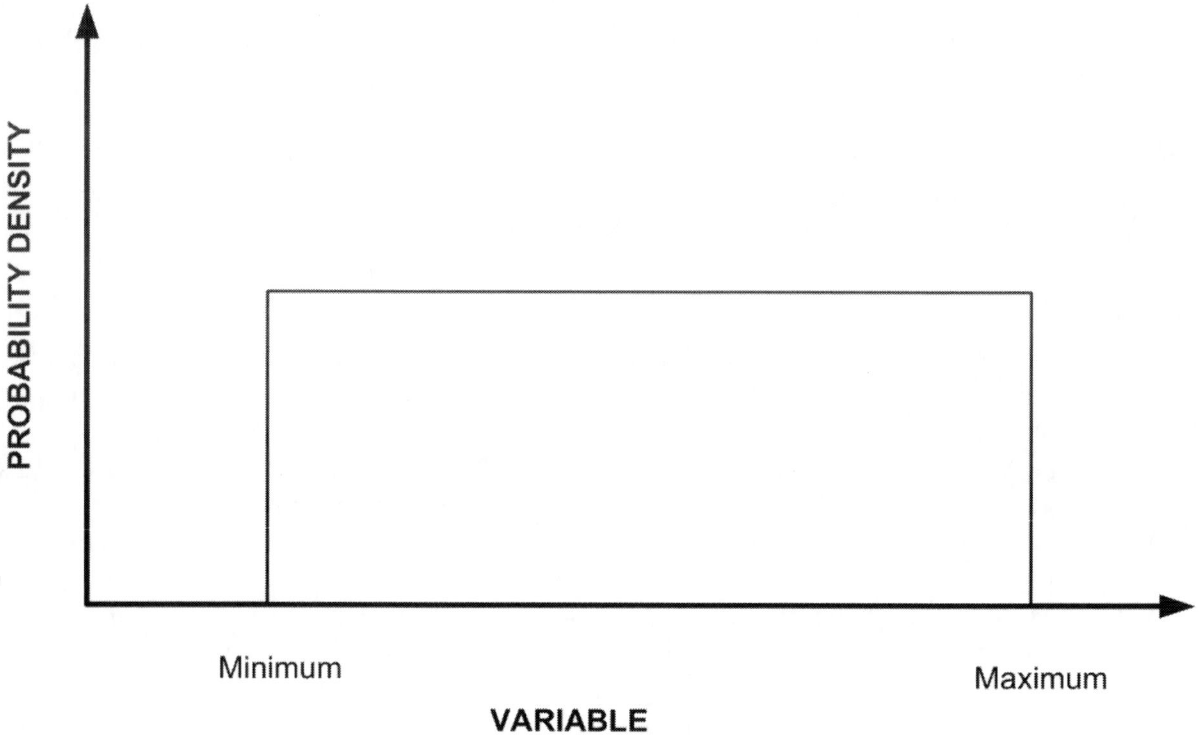

Figure A-1. Uniform Probability Distribution

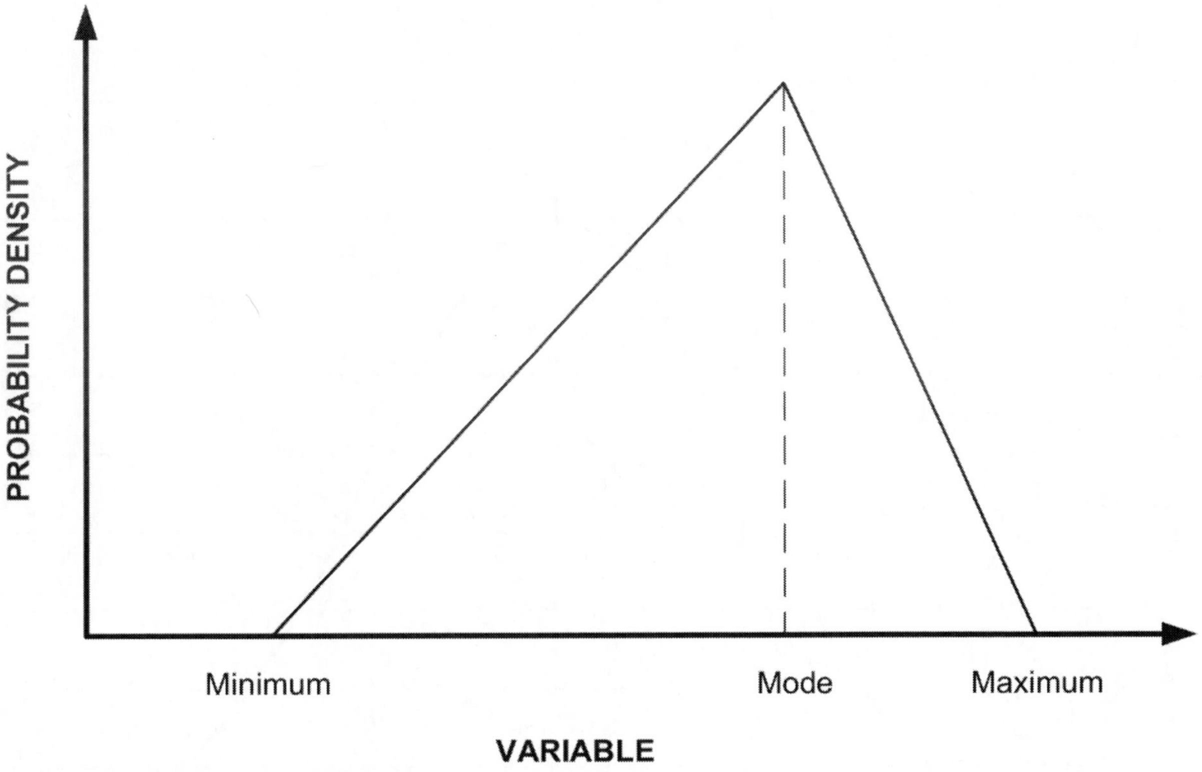

Figure A-2. Triangular Probability Distribution

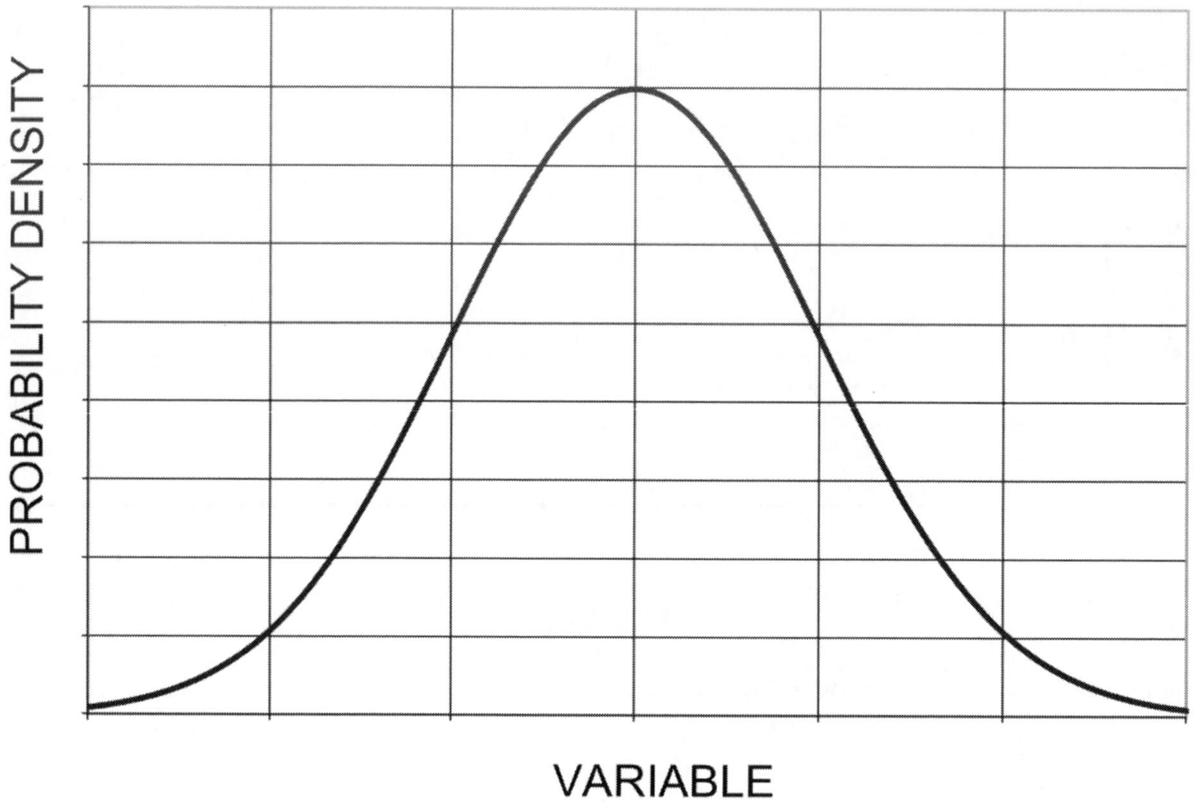

Figure A-3. Normal Probability Distribution

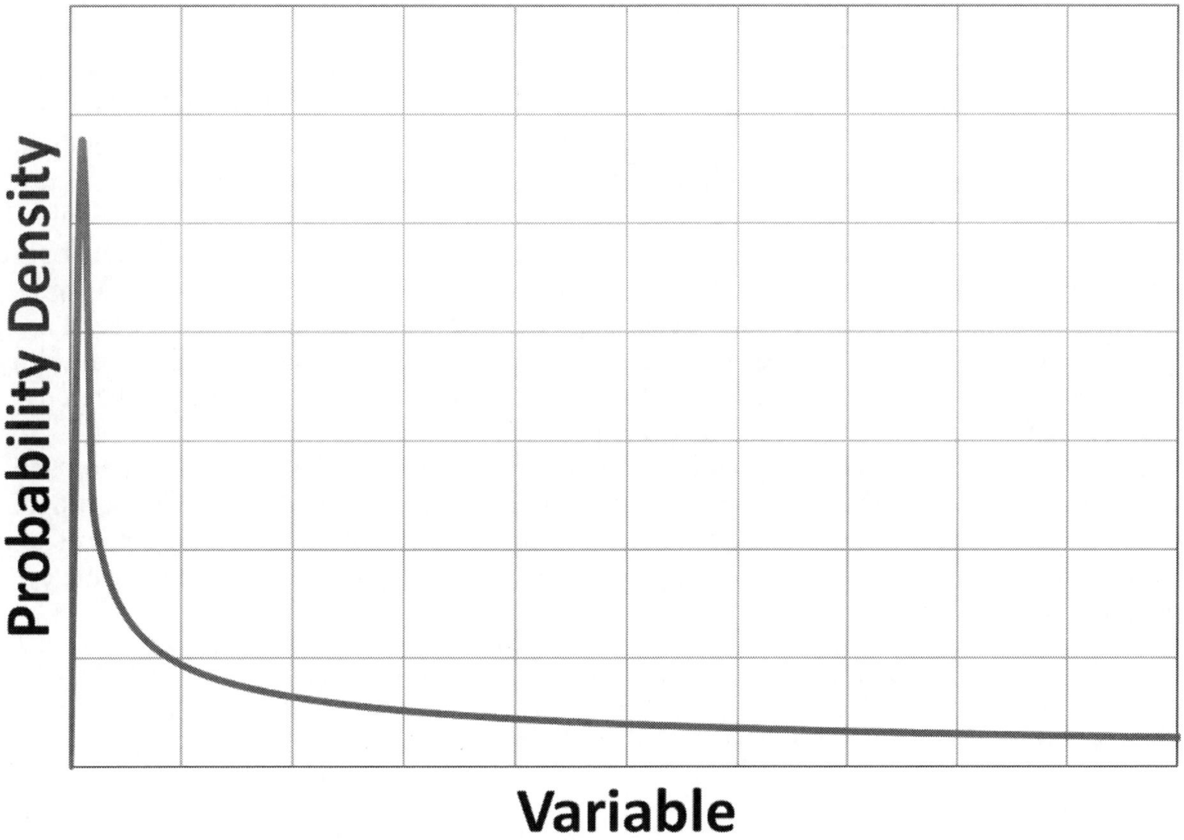

Figure A-4. Lognormal Probability Distribution

590424 Making the Most of Your PHA

Abstract:

A Process Hazard Analysis (PHA) is required to address OSHA Process Safety Management (PSM) and EPA Risk Management Program (RMP) regulations. Understanding the potential process hazard is widely recognized as a foundational element of PSM and RMP, and in addition to simply meeting regulatory requirements, can be a critical mechanism for flushing out potential vulnerabilities early in the design process for capital projects. Although an important activity, PHA is also recognized as a sizeable investment in time and energy for an Owner/Operator. So why not make the most of it? Other important activities at plants such as Mechanical Integrity, Alarm Rationalization, and Management of Change are able to stay focused and possibly be streamlined (i.e., reduced cost) based on the ability to use the results of a quality PHA having the necessary information and usability of documentation. So why not structure the PHA and documentation to be able to capitalize on potentially significant cost-savings and possible program integration?
The thrust of this paper and conference presentation will be to identify tips for conducting and documenting the PHA to maximize its ability to be used for supporting other important programs such as MOC, MI, and Alarm Rationalization. The objective will be to allow PHA Coordinators to be able to readily apply these tips to bring about improvements and cost-savings on these interrelated programs.

589814 Theoretical Analysis of the Viability of Isopropanol-Cyclohexane Solvent Systems for Biomatter Lipid Extraction

Abstract:

A theoretical alternative solvent combination for the extraction of lipids from biomatter is the Water, Isopropanol, and Cyclohexane instead of the more common combination used in the Bligh and Dyer Method (Water, Methanol, and Chloroform). While the Bligh and Dyer Method is considered effective, toxicological concerns have been raised over its use of both Chloroform and Methanol. The alternative method utilizes solvents that do not pose as high as a toxicological concern while theoretically maintaining a similar level of practical viability. The alternative method's extraction efficacy was judged against the performance of the Bligh and Dyer method using the simulation software Aspen Plus V10®. The equilibrium data generated from Aspen Plus V10® was then graphed into tertiary diagrams for data analysis using the graphing software SigmaPlot.

592166 Techno-Economic Analysis of the Conversion of CH4 and CO2 Intoacetic Acid Using Microwave Plasma

Abstract:

Methane and carbon dioxide together represent 90% of greenhouse gases [1], and one way of diminishing their emissions is through the development of pathways that convert them into added value chemicals. Acetic acid is one of such chemicals, and efforts at the laboratory scale have been carried out to obtain alternative ways of producing it. However, there are very few studies that evaluate the possibility of implementing these processes at a larger scale. In this paper, techno-economic and sustainability analyses were used to assess the feasibility of upscaling two pathways to produce acetic acid: (1) by direct conversion of methane and carbon dioxide and (2) by indirect conversion, using syngas (obtained from dry methane reforming) as an intermediate. For the direct method a microwave non-catalytic plasma reactor was considered as the main driver in the process simulation [2]. For the indirect method, a zeolite catalytical single step reaction unit was considered [3]. APEN Plus was used to simulate the reaction and purification systems of both methods, and alongside literature values and plant cost estimation techniques, the capital and operational expenses were calculated. Finally, the feasibility of both projects is assessed on carbon sequestration mass balances and economic metrics. The main sensitivity variables that need to be improved in order to obtain a minimum ROI threshold value were also evaluated.

Key words: carbon management, process design, microwave plasma reactor

[1] EPA. *Overview of Greenhouse Gases.* United States Environmental Protection Agency. 2017.

[2] Wang, L., Yi, Y., Wu, C., Guo, H., Tu, X. *One-Step Reforming of CO2 and CH4 into High-Value Liquid Chemicals and Fuels at Room Temperature by Plasma-Driven Catalysis.* Angewandte Chemie. 2017.

[3] Lopez, L., Velasco, J., Montes, V., Marinas, A., Cabrera, S., Boutonnet, M., Jaras, S. *Synthesis of Ethanol from Syngas over Rh/MCM-41 Catalyst: Effect of Water on Product Selectivity.* Catalysts. 2015.

589668 Model Predictive Control Design for the Allam Power Plant Integrated with an Air Separation Unit

Abstract:

Since electricity and heat generation account for around 25% of this greenhouse gas emission, high efficiency power generation technology such as the Allam cycle which emits essentially no CO_2 and NOx is much needed. Being an oxy-fuel power cycle, the Allam power plant is inherently integrated with an air separation plant (ASU). In this study Aspen Plus Dynamics was used to model the integrated Allam power plant/Air Separation Unit (ASU) with a high degree of heat and work integration based on the steady state work of Net Power. In our previous work, regulatory and advanced PID controls were implemented and tested for various set point tracking and disturbance rejection scenarios. However, PID controllers have limitations for various reasons. Consequently, a model predictive control (MPC) was designed in this study to optimize and improve the performance of the integrated unit and to test its performance in terms of dynamic behavior and robustness to model uncertainties.

589672 Explore the Function of Ion-Exchange Membrane (IEM) in Membrane Capacitive Deionization (mCDI) Cell, an Alternative Technique in Brackish Water Desalination Field

Abstract:

In the arid west, the freshwater supply of many communities is limited, leading to increased interest in tapping brackish water resources. Although reverse osmosis is the most common technology to upgrade saline waters, there is also interest in developing and improving alternative technologies. Here we focus on Capacitive Deionization (CDI), which has attracted broad attention as an energy efficient desalination technology, especially for ion extraction from low salinity brackish water. The objective of this work is to explore the function of ion-exchange membrane (IEM) in enhancing the performance of CDI cell. In this study, a fully coupled two-dimensional model describing ion transport and adsorption behavior in Membrane Capacitive Deionization (mCDI) cell was developed and solved numerically, taking advantage of Darcy's Law, Nernst-Planck equation, and Donnan equilibrium considering hydraulic dispersion in channel, non-ideal membrane, and modification of ion diffusivity in porous media. The model behavior was supported by the experimentally observed variations in transient effluent concentration under different operating conditions. The results showed that ion flux was significantly improved because of the IEM, leading to a faster desalination rate. The equilibrium adsorption amount is increased due to the permselectivity of the IEM, resulting in extra ion accumulation in electrode's macropores during desalination process. The cut-off time (optimal time to switch polarity to generate highest energy efficiency) of mCDI was shorter with a higher salt removal percentage compared to CDI under the same operating conditions. A sensitivity study of the thickness and the fixed charge density (starting from 500mM) of the IEM was done separately, and neither of them were found to be the dominant influencing factors under a low concentration range of brackish water (below 100mM) in this study.

590103 An Integrated Process for the Production of Propylene Glycol from Shale Gas

Abstract:

Propylene glycol (PG) is produced on a large scale primarily for the production of environmentally safest and least toxic industrial and domestic products such as polyester resins, thermoset plastics, polyurethanes, food additives and antifreeze. In this research, a novel integrated plant is developed for the production of PG from shale gas. This integrated approach has the benefit of safer operating conditions since the intermediate propylene oxide, which is explosive, does not need to be stored and transported. Furthermore, there are potential economic benefits from integration. The overall plant is simulated in the Aspenplus environment, and a variety of process conditions are tested at steady-state to optimize the production of PG. Heat-integration tools are utilized for energy-saving and capital cost reduction opportunities. A comparative economic assessment based on existing plant information indicates that the use of process integration techniques has the potential to reduce costs significantly.

Keywords: Process Integration, Heat Integration, Propylene Glycol

592456 Monitoring Methods for Fault Detection and Diagnosis: Algorithms and Applications

Abstract:

Many industrial processes collect an abundance of data from different sensors. These sensors often measure a wide variety of physical properties in order to ensure that these parameters are adhering to expected values. This is essential to ensure plant and operator safety, increase economic benefits, and maintain product quality.

A number of algorithms have been developed in order to improve existing fault detection and diagnosis performance. These algorithms integrate a number of different data-driven driven tools and methods. Multiscale wavelet-based representation of data can be used in order to handle data that is autocorrelated, non-Gaussian, and noisy. Hypothesis testing methods such as the Generalized Likelihood Ratio (GLR) technique can be used in order to provide the best possible detection for a fixed false alarm rate. Moreover, certain model-based methods have been developed in order to monitor process drifts and degradations in the process model, even when a process is operating under control.

This work will discuss the different tools utilized in development of the algorithms, their features, and practical applications, including the benchmark Tennessee Eastman Process (TEP).

589902 Selective Separation of Industrial Relevant Metal Ions from High Tds Water Using Wafer-Enhanced Electrodeionization (WE-EDI)

Abstract:

There is a need for an efficient and economical process of ion-selective separation. In hydraulic fracturing, for instance, several million gallons of water are used in a site, most of which are disposed of by deep well injection taking the water from the ecosystem and causing earthquakes. However, due to the high amount of ions in wells at fracking sites, pressure driven processes for water recovery are impossible to use. This project will demonstrate an emerging environmentally friendly Wafer Enhanced Electrodeionization (WE-EDI) technology which attempts to eliminate the shortcomings that come with normal EDI and separates effluent streams into concentrated ions and clean water. In WE-EDI, ion-exchange resins are replaced by wafers that consist of cation and anion exchange beads and a polymer binding agent. In this study, selective separation of these ion exchange wafers and the effect of electroosmotic drag on the membrane structure have been performed for various anions and cations (Na+, K+, Ca2+, Cl-) and improved selectivity and higher water recoveries have been observed while significant modeling has taken place to predict the selective separation as a strong function of the equilibrium constants between the ions and the beads. This project is supported by the Membrane Science and Technology (MAST) Center and is conducted with the collaboration and guidance of industry mentors.

593303 Influence of Particles on Amine Losses during CO_2 Capture: A Process Simulation Coupled Aerosol Dynamics Model _

Abstract:

Influence of Particles on Amine Losses during CO_2 Capture: A Process Simulation Coupled Aerosol Dynamics Model

David Dhanraj and Pratim Biswas

Aerosol and Air Quality Research Laboratory

Department of Energy, Environmental and Chemical Engineering

Washington University in St. Louis

St. Louis, MO 63130, USA

Efforts to reduce the concentration of atmospheric CO_2 have been initiated due to escalating concerns of global warming and climate change. Amine-based post-combustion CO_2 capture offers an economically viable retrofit for power-plants in combating these emissions. However, solvent losses have been a major challenge to commercializing this technology. Recent experiments[1] show that vapor-based and aerosol-based solvent losses are the key mechanisms of amine solvent loss during CO_2 capture. While vapor-based amine losses can be abated by employing techniques like water-washing, there has been little work to curtail aerosol-based solvent losses. This is primarily because the underlying mechanisms governing these losses are still not well understood. Although there are a few numerical models proposed for this system, there are no models that consider aerosol phenomena such as coagulation or particle loss. Moreover, these models have not been fully validated through experiments.

Furthermore, there are a few techniques that are being developed to tackle solvent losses. It is known that reducing particle number concentration in the flue gas will reduce amine emissions. Conventional Electrostatic Precipitators (ESP) are however inefficient in capturing particles in the size range 0.1 – 1.0 µm. Amongst different approaches demonstrated, introducing a pre-charging stage and a photo-ionization stage has shown potential in increasing the charging efficiency.

Therefore, the broad objectives of this work are primarily focused on two facets of the amine loss problem. The first is the development of a model that integrates aerosol phenomena with a process engineering model that accounts for the reactive mass transfer process in a packed-bed absorber. The second is the design, fabrication and testing of a bench-scale Photoionization Enhanced Two-staged Electrostatic Precipitator (PI-ESP) at Abbott Power Plant. In essence, with the PI-ESP aimed at reducing the particulate concentration in the flue gas, and the model relating the influence of particulate concentration in the flue gas to amine emissions, this work aims to illuminate the underlying mechanisms governing amine emissions, and demonstrate a solution to tackle the problem.

References

1. International Journal of Greenhouse Gas Control, vol. 85, pp. 86-99, 2019.
2. International Journal of Greenhouse Gas Control, vol. 19, pp. 138-144, 2013.

590033 Can Transient Data Help to Speed up Kinetic Model Construction in Hydroprocessing?

Abstract:

INTRODUCTION

Kinetic model is a powerful tool for the adequate design and simulation of chemical processes. It is usually developed based on the steady-state experimental data. Kinetic modeling of hydroprocessing processes is a difficult task since the crude oils contain many compounds with complicated structures [1]. The long stabilization in hydroprocessing (several days) leads to long experimental campaigns to acquire sufficient steady-state data [2]. However, in the transient regime towards the steady state, effluent analyses are already carried out at regular time intervals to verify whether the steady state has effectively been reached. These transient data are available but are currently not used for kinetic modeling because the stabilization phenomenon is not taken into account in the model. In this respect, the present work aims to speed up the kinetic model construction by employing these transient data. It helps to reduce the number of experiments for kinetic modeling and to obtain a model more robust to measurement uncertainty. The experimental design technique is also studied in the work to select the best representative operating conditions for model fitting.

MATERIALS & METHODS

A transient model with consideration of the stabilization based on time on stream is developed. It is compared to the classical steady-state kinetic model to demonstrate the advantages of exploiting transient data.

Experimental data for modeling cover a variety of feedstocks over a hydrotreating catalyst with a wide range of operating conditions: total pressure between 90 and 140 bar, liquid hourly space velocity (LHSV) between 1 and 3 h^{-1} and temperature between 370 and 400 °C. A series of consecutive experimental points with the same operating conditions leading to the steady state point is denoted as an ‘episode’. One experimental test corresponds to approximately 7 steady-state points, i.e, 7 consecutive episodes. The latter correspond to around 50 experimental points (7 steady-state points and 43 transient points).

The whole database is divided into the calibration and validation database via the Kennard-Stone algorithm [3,4]. Based on the Euclidean distance, Kennard-Stone technique intends to select the best representative subset of all the candidates available in the entire database. Here, each episode is considered as a candidate.

The impact of the calibration database size is also investigated. The comparison of transient model and the classical steady-state model was carried out several times via different calibration and validation databases size.

Furthermore, in order to test the robustness of transient model to the uncertainty of measurement, noise is artificially added to a certain number of points in the calibration database by varying them in a random manner. The “noisy” calibration database is then used to train the models and then test on validation database.

RESULTS & DISCUSSIONS

The results indicate that the transient model can be developed from a lower number of experiments thanks to using transient data. It is competitive to classical steady-state kinetic model. Transient model is also more robust to measurement uncertainty. Kennard-Stone technique has been found to provide a good selection of operating conditions for the design of transient experiments. This technique can also be applied for complex models when no initial parameter guesses available.

The proposed methodology provides an intriguing potential of saving tremendous time and cost in experimental work for kinetic modeling and it can be applied to the reactions where the chemical stabilization requires significantly more time than the analysis time.

References

[1] A. T. Jarullah, I. M. Mujtaba, and A. S. Wood, "Kinetic model development and simulation of simultaneous hydrodenitrogenation and hydrodemetallization of crude oil in trickle bed reactor," *Fuel*, vol. 90, no. 6, pp. 2165–2181, 2011.

[2] N. Y.P. Cao, B. Celse, D. Guillaume et al., "Assessment of the dominant factors during hydroprocessing stabilization," in *Computer Aided Chemical Engineering*, vol. 46, pp. 1129–1134, 2019.

[3] A. Saptoro, M. O. Tadé, and H. Vuthaluru, "A modified Kennard-Stone algorithm for optimal division of data for developing artificial neural Network Models," *Chemical Product and Process Modeling*, vol. 7, no. 1, 2012.

[4] R. W. Kennard and L. A. Stone, "Computer aided design of experiments," *Technometrics*, vol. 11, no. 1, pp. 137–148, 1969.

589562 CFD Simulation Analysis for Refinery Plot Plan Design Verification

Abstract:

With increased computational capability, simulation of refinery scale physical processes have been made possible. In this presentation, we will discuss few selected works from our recent plot plan design verification based on the results of Computational Fluid Dynamics (CFD) simulation analysis. The examples shall include heat effect study due to relative position of Heaters and Coolers, Sulphur dust dispersion in Sulphur Handling Facilities, and water droplet spill from Cooling Tower. In all of these cases, CFD simulation quantified the effect of one unit on the adjacent unit due to their relative position and helped assist either in the justification of the layout design or design modification. A general guideline on the relative positioning of key processing unit as relevant to a modern refinery and petrochemical complex will also be proposed.

588983 Renewable Energy in an Oil Exporting Region

Abstract:

In a region having one quarter of the world known reserves of oil and gas, does it make sense to talk about renewables? In this paper we will discuss this subject and explore current electricity demand and future projections and if renewable are competitive position in this region, what kind of renewable energies are suitable for the region and for specific applications, basics of solar energy and go over some success stories in the region.

Revamp Trends, Strategies, and Challenges for CDU/VDU and FCC Units

Avinashkumar V Karre, Worley Group, Baton Rouge, LA 70817

Abstract:

Upgrading or revamping existing units in the refinery has been a trend followed by many operating companies in recent years. This is due to high profitability gains mainly by utilizing majority of the existing infrastructure and by expanding processing capability of various crudes available in the market. Most of the projects are mainly capacity expansion driven and few are due to environmental regulation mandatory requirements. Most of the projects which are driven by capacity expansion also address existing processing unit operational problems such as inadequate relief disposal system. Refinery revamp projects are inherently challenging due to the several limitations imposed by existing equipment design, existing utility capacities, turnaround timing, and plot space.

Capacity expansion of crude and vacuum units is very commonly done as different crude types are economically available in the market. Many refiners are experiencing unexpected shutdowns routinely due to scaling and fouling in process equipment. Run length of 8 years between scheduled turnarounds is expected from any operating company. Key to avoid fouling and coking is by maintaining adequate velocities in heaters and heat exchangers. Replacement of heat exchangers, trays replacements, burner replacement in heaters, pump upgrades, and ejector system upgrade for higher vacuum are common upgrades. All crude and vacuum units start to perform poorly with time due to scaling and coking. Following systems are commonly affected by high run length - desalting, preheat trains, crude and vacuum tower overhead system fouling and corrosion, low diesel and vacuum gas oil (VGO) recovery, and heater coking. Relief and disposal system are often evaluated after a major revamp.

Capacity expansion of Fluid Catalytic Cracking unit (FCCU) is also commonly done which are driven by product volumes. Revamp jobs include replacement or upgrades to blower, regenerator cyclones, feed distributor, fractionator trays with packing, fractionator heat exchangers, wet gas compressor. Reactor head replacement is due to corrosion.

1. Introduction:

Crude distillation unit (CDU): CDU unit consists of raw crude heated in several heat exchangers before it's desalted in desalters. Desalted crude is further preheated in several heat exchangers before flashing light streams in Preflash drum or tower. Preflashed crude is preheated in several exchangers before it's heated in a fired furnace. Heated preflashed crude is sent to the bottom of the crude column and several components such as off-gas, light straight run naphtha, heavy straight run naphtha, kerosene, diesel, heavy gas oil, and reduced crude are separated at different stages and temperatures. Side stripper columns are often used for the products streams such as kerosene to remove any entrained gas components and maximize the yield. Several pump arounds are often added around the crude column to provide cooling duty. Pump arounds and product streams are cooled by heating the raw crude, desalted crude, and preflashed crude as mentioned earlier depending on the temperature driving force and heat capacities. Pump arounds, final products, and intermediate products are pumps to provide desired pressures in the system.

Vacuum distillation unit (VDU): Separated reduced crude from the crude column is further preheated in several exchangers before feeding into vacuum furnace. The heated reduced crude is fed to vacuum column where its separated into off-gas, low vacuum gas oil, medium vacuum gas oil, heavy vacuum gas oil, and resid or vacuum bottoms. A desired vacuum is maintained on the column using 150 psig steam ejectors followed by condensers and hot oil well. Several pump arounds are added around the column to provide a necessary cooling duty.

Fluid catalytic cracking unit (FCCU): All heavy to medium gas oils are feed to the FCCU. Combined feed is preheated before feeding to the riser tube of the reactor. FCCU catalyst cracks and isomerizes to useful and transportable fuels. Coked FCCU catalyst is burned with air in the regenerator. Cyclones separate flue gas from the solid FCCU catalyst. Fresh FCCU catalyst is then recycled back to the reactor rise tube for further reaction. The gaseous product formed in the overhead section of reactor is feed to the fractionator column which further separates into fractions such as off-gas, naphtha, kerosene, diesel, and unconverted oil. Several pump arounds and coolers are used to balance heat duty around the column.

2. Revamp strategies:

Several parameters affect the revamp operation of a CDU/VDU and FCCU. Such parameters are discussed in detail as below -

2.1 Type of crude available

Type of crude in terms of its light fractions, sulfur content, API gravity, and available salt content are some of the key parameters each refinery is looking for. Heavy crude is cheaper compared to lighter crude but is difficult to process. Fractions separated from lighter crude are more valuable compared with heavy crude. Higher sulfur content in the crude means more corrosion in the equipment and more downstream processing is required such as merox unit or amine treatment unit. Higher salt quantity means the desalter sizing should be adequate to handle additional salt loading. If the crude has higher quantity of gas oils, one could use that to feed to

FCC to produce additional higher value products provided FCCU has the additional processing capacity.

2.2 Flow parameter

More feed means higher velocities through pipes and heat exchanger tubes. Also, higher flow demands higher tray capacities, higher reflux, and higher pumparound rates. For revamp operations, each exchanger, pump, column, trays, control valves, and drums are checked against the original design capacities. Oftentimes, pumps are upgraded, higher capacity trays are added, and parallel heat exchanger train is added. CDU and VDU furnaces are often modified to provide higher heat input required for the higher throughput.

Same is applicable for FCCU regarding heat exchangers, pump capacity, and tray capacities. Moreover, cyclones on the product and regenerators are replaced for the higher throughput.

2.3 Type of refinery

Each refinery is originally designed to make certain products depending on light or heavy or blend of crudes. If the refinery is not designed to process heavy crude, it is limited to light crude and revamp is focused more on light ends such as off-gas and naphtha. FCCU plays a crucial role in maximizing naphtha yields if the feed to the CDU/VDU is enriched in gas oils.

2.4 Sulfur in crude

Sour crudes invite more maintenance problems and needs finishing of products and present a challenge in the treating off-gases. Crude column overhead and FCCU reactor overhead are often replaced if the sulfur content in the feed stream is increasing. The off-gas streams from CDU and FCCU needs further sulfur treatment before blending into the fuel gas header.

2.5 Energy requirement

Higher feed rate requirement needs higher preheat and higher vapor lift-up in the columns. Additional parallel preheat trains are added to capture the heat and that also helps in reducing pressure loss through the preheat train. Fired heaters are revamped with higher duty by either replacing burners and/or by adding additional air into the system.

2.6 Cooling duty requirement

By adding additional heat into the system for higher throughput, additional heat needs to be removed from the columns. This is achieved by energy integration around all the pumparounds and product streams. For higher efficiency of energy recovery, energy integration between multiple units is often done by sacrificing a flexibility in operating the units.

2.7 Utility requirement

Higher cooling and higher heating demand need additional utilities such as boiler feed water, fuel gas, fuel oil, steam, cooling water, etc. For revamp projects, dependence on cooling water is generally not preferred to avoid additional load on existing cooling tower or new cooling tower. Heat integration close to 5 to 10°F temperature approach is often done, and air-cooled fin fan design is preferred.

2.8 Desired product

Each refinery is designed to optimize LPG, naphtha, kerosene, gasoline, and diesel products. Heavy to medium gas oils are used in navy and agriculture. CDU, VDU, and FCCU are optimized accordingly depending on the feed and the product desired. For e.g. if the desired product is heavy gas oil, partial revamp of VDU unit is necessary to attain higher boiling temperature by reducing vacuum in the column. Ejectors and overhead condensers system of the VDU unit is replaced for altering vacuum in the column.

3. Revamp trends:

3.1 Environmental regulations

CDU and VDU do not have any environmental regulations with processing higher sulfur crude, but the off-gas, kerosene, gasoline, heavy fuel oil, and LPG products must meet certain sulfur limits before selling to end customers. So, processing of higher sulfur crude impacts the downstream finishing units such as kerosene merox, LPG sweeting etc.

3.2 Capacity debottlenecks

Each refinery likes to push the limits of their existing equipments with minimal changes and lowering capital costs if possible. Also, increasing throughput through the units is very common practice to maximize product yields and gain higher refining margins to meet corporate goals. Detailed economic analysis of a revamp project is studied against the profits gained over time. If feasible, capacity debottlenecking of the existing unit utilizing existing equipments as much as possible is done in a desired timeframe.

4. Challenges in revamp:

Revamp of units is often very complex and challenging task. Some of the key challenges are listed below.

4.1 Turnaround timing

The revamp project must account for turnaround timing for the units. Early tie-in work are often done whenever appropriate turnaround window is available for the units. Relief systems, fire related equipments, and utility systems are most critical when it comes to revamp projects. Utility systems are very critical and needed by the entire refinery, for these reasons, tie-ins are often done using special welding technique while utility fluids are flowing through the pipe.

4.2 Relief system limitations

With a higher throughput requirement, higher relief load from the units is expected. A detailed relief system study is carried out. For revamp projects, alternatives to modifying the flare system such as adding high critical safety instruments are often considered to reduce costs and depending on local regulations.

4.3 Plot space limitation

Plot space limitation is often a challenge for any revamp jobs. Finding an alternate plot space location to minimize piping if feasible is carried out. Multiple or stacked floors are often considered to reduce cost. An alternative to shell and tube heat exchanger such as plate and frame heat exchangers are considered when plot space is very critical. Plate and frame heat exchanger cannot be used for all fluid types but only for clean fluids such as LPG. Fin-fan are also often considered on top of pipe rack as an alternative to shell and tube if feasible. The size of the drums are optimized and reduced by adding internals which increases the separation efficiency.

4.4 Increased demand for renewable energy/fuel

There is an increased demand in renewable energy such as biodiesel due to increased demand of electric cars in the market, and traditional liquid transportation fuels such as gasoline will phase out eventually by 2040. Due to reduced future demand in gasoline, every refinery will be trying to add renewable projects to their profile to diversify and be competitive in the global market.

5. Conclusion:

Revamp of CDU/VDU and FCC driven mainly by the capacity expansion agenda to maximize refining margins and maximize utilization of existing equipments as much as possible. Revamp of these units includes addition of preheat trains, upgrading pumps, replacing trays, upgrading furnaces, replacement of ejector system, energy integration of pumparounds and products, evaluation of existing instruments, replacement of cyclones, replacement of crude and FCC overhead sections, upgrades to relief system, and turnaround work. Revamp of these units is very challenging compared to just adding a grass-root unit because of challenges such as limitations of plot space, turnaround windows, limitations of existing equipments and relief system, and rising demand in renewable energy.

589585 Light Naphtha Isomerization: An Innovative 2 in 1 Compact Revamp

Abstract:

The once-through conversion of a traditional fixed bed light naphtha isomerization unit is limited by the reaction equilibrium. Overall unit conversion can be achieved by use of a recycle stream of primarily n-paraffins, a typical example of which is the use of a de-isohexanizer (DIH) column to separate the recycle stream as an n-hexane rich middle cut, with the overhead isomerate and bottom product “drag stream” sent to gasoline blending. Even the simple addition of a DIH is capital and energy intensive and the isomerate product quality is limited to around 88 RON. Any incremental feedstock processing demands place stress on the system resulting in lower isomerate octane.

KBR’s MAX-ISOM™ technology uses a unique catalytic distillation column design to generate high conversion of n-paraffins to iso-paraffins in a single column. While the technology can operate as a stand-alone isomerization unit in its own right, it also offers a unique revamp opportunity to already constrained recycle isomerization units. MAX-ISOM can be placed downstream from the DIH column of a traditional recycle isomerization unit to process the middle and bottom products. The DIH column becomes a binary split column, thereby reducing energy consumption, and the volume of fresh feed to the existing unit can be increased by up to 40% due to removal of the recycle stream. Furthermore, overall isomerate octane will be boosted due to high conversion achieved in the MAX-ISOM column.

The MAX-ISOM column contains beds of isomerization catalyst separated by fractionation zones. High conversion of n-hexane and methylpentanes is achieved by exploiting the relative volatilities of the reactants and products and through internal recycles. This paper will include a case study to demonstrate the throughput and product quality gains offered by MAX-ISOM in a revamp flow scheme.

590689 Oilfield Water Management for Sustainable Unconventional Shale Development: Case Histories and Lessons Learnt

Abstract:

It is estimated that United States will surpass Saudi Arabia as the major exporter of oil and liquids. Multi well pad drilling and multi stage hydraulic fracturing has dramatically increased hydrocarbon production in United States. Water is a critical component of drilling, completion and production operations in oil and gas. In order to sustain the current rate of unconventional growth for a cash crunched sector, its imperative to recycle and reuse produced and flowback waters, thereby decreasing the fresh water demand for hydraulic fracturing operation. Proper knowledge and management of oilfield water life cycle can directly impact operational efficiencies and reduce operating costs. Understanding the impact of critical components in oilfield water on fracturing fluids and the reservoir, can enable operators to minimally treat produced and flowback water for reuse.

The key is to take a holistic approach in-terms of technology evaluation, application, treatment, logistics and waste reduction for an economically feasible water management solution. The study will focus on produced water constitution, its impact on reservoir, key technologies used in produced water recycling and treatment, the end result with case histories showing almost 135MM Bbls of water treatment and reuse over last few years.

583094 Mild Thermal Cracking of Oil Sands Bitumen

Abstract:

A significant portion of Canadian bitumen is transported via pipelines by blending the high-viscosity bitumen with diluent at a ratio of 25-50 vol% to meet pipeline specifications for viscosity (350 cSt at 7°C) and density (940 kg/m^3 at 15.6°C). In addition to being expensive, diluents occupy valuable pipeline capacity and must be separated in refineries only to be transported back to bitumen production sites for reuse. Raw bitumen is sold at a discount price on global markets and has limited access to current and new markets due to its low quality. A number of recent studies have shown that partial upgrading technologies, especially if integrated with upstream operations, can effectively overcome these challenges to reduce, or even eliminate the use of costly diluent, increase bitumen quality and market value, potentially decrease GHG emissions, and reduce capital and operating costs compared to bitumen full upgrading. Currently, most of the partial upgrading technologies under development are based on thermal processing. However, due to some undesired properties of thermally cracked products, such as incompatibility and high olefin content, commercial visbreaking processes cannot convert bitumen directly into a product that meets pipeline specifications. In this study, we focused on the investigation of product qualities at different thermal cracking severities to achieve maximum reduction of density, viscosity and total acid number (TAN), while ensuring reasonable product compatibility and stability.

589587 K-Saat, a Break-through Solid Acid Alkylation Technology

Abstract:

Globally, the demand for cleaner burning gasoline blend components is increasing while overall demand for gasoline is declining. The increase in demand for cleaner burning gasoline is driven primarily by environmental mandates and high performance engines. Alkylate, isomerate, ethanol and ethers represent the cleaner burning gasoline blend components. Among these, alkylate is the ideal clean fuel component because it has a high octane rating, exhibits low vapor pressure, low sulfur levels, no aromatics or olefins. Alkylate is currently produced using either sulfuric or hydrofluoric acid as the catalyst. Both acids are dangerous and can lead to corrosion-induced accidents and impede firefighting efforts. Solid acids have for many years promised safer and cleaner alkylation. The poor stability of most solid acids, however, have resulted in expensive processes with complex reactors and large catalyst inventories, making them uncompetitive with liquid acid technologies.

After years of development, a new solid-acid catalyst technology ExSact has reached the point of outperforming these liquid acids. The engineered solid-acid catalyst has been designed to give over 24-hr cycle-times, robust resistance to typical poisons, the ability to handle a variety of feedstocks (ethylene to amylene), a simple adiabatic fixed-bed reactor design and regeneration using hydrogen while increasing he overall octane pool. The ExSact catalyst forms the core of an inherently safe and efficient K-SAAT process that generates high-octane alkylate. K-SAAT reactors are the best option available to refiners worldwide to move away from HF Alkylation, offering a commercially proven process with high return on investment and vastly diminishing the risk profile. This is achieved by replacing/debottlenecking existing liquid acid alkylation reactors and reusing the existing recovery system (DIB/Debutanizer/Depropanizer/Isostripper).

The stable catalyst performance greatly simplifies the overall process design, which reduces the capital cost of the alkylation plant, while lowering energy consumption. KBR has licensed K-SAAT units and will discuss overall commercialization status in addition to performance data from the operating unit.

589404 Fired Heater Convection Section Steam Generation Tube Failure (*Understanding the impact of flow regime and high peak heat flux profiles)*

Abstract:

Two catalytic reformer unit heaters were facing convection section steam generation tube failures recurring every year. This problem was determined by low-pressure drop across steam generation banks and subsequent IR scanning of tubes. It demanded frequent removal of failed tubes. Initially, caustic gouging, poor tube side flow regime, high localized heat fluxes and improper flue gas flow distribution were thought as a possible reason for tube failure. However, water chemistry was being looked at and was not a contributing factor for the tube failures as other heater receiving the same boiler feed water was not facing tube failure issues.

Further, CFD modeling was adopted to study the flue gas flow distribution between the heaters. A damper was installed in the flue gas duct of the heater to control the flue gas flow distribution. Thus, high peak heat flux and poor tube side flow regimes were identified as the major factors causing tube failures.

- The heaters were operating with circulation ratio in the range 7 to 9 and the peak heat flux in the tube failure region was also very high (25,000 – 50,000 Btu/hr-ft^2).
- An unstable slug flow regime was predominating in the tube failure region of both the heaters.

This paper describes the scheme to overcome these issues and eradicate tube failures.

589927 A Simulation Workflow for Flow, Thermal, and Vibration Analysis of Hvu Transfer Line

Abstract:

In High Vacuum Unit (HVU), vacuum gas oil is recovered from long residue feed. Vacuum gas oil yield, reliability and operability depend on correct design of the vacuum transfer line. Correct and safe design of Transfer Line also prevent excessive cracking and enable a longer run length before decoking. A simulation model is developed for analysis of HVU transfer line flow velocity and pressure drop profile considering flashing along the line. Cases of both normal operation and start-up conditions are considered. Subsequently, a thermal and a vibration analysis model were also developed to look into the thermal stress and vibration safety of the design. In this presentation, we will discuss the complete workflow of simulation analysis highlighting the critical aspects in the correct and safe design of HVU Transfer Line.

589974 Systematic Synthesis of Wastewater Treatment Networks Using the P-Graph Approach

Abstract:

Wastewater treatment is an area of utmost importance to address the challenge of decreasing freshwater resources. Traditionally, wastewater treatment consists of three or four sequential stages: preliminary, primary, secondary and tertiary [1]. Preliminary treatment is intended to facilitate the removal of contaminants in the following stages. Primary treatment mainly involves the withdrawal of settleable solids. The secondary treatment is capable of removing pollutants such as chemicals and metals. In some cases, a tertiary treatment might be used for further purification if the required purity limits are not achieved during the first three stages. To create this systematic approach for designing wastewater treatment networks [2], we rely on the principles of mathematical modeling to generate an exhaustive enumeration of all the possible technologies and their connections during the early stages of design. This enables the generation of a maximal structure which consists of all possible treatment technologies, and stream connections. Some of these structures would be non-intuitive and could include recycling, reprocessing, and multiple technologies in parallel or series to remove the same contaminant, which would not have been obtained via typical design heuristics [3]. Thus, the sequence of treatment technologies predicted using this maximal structure would be based on holistic consideration and would not lead to suboptimal solutions.

Representative case study of a wastewater stream containing settleable solids and chemicals is considered as the motivating example where the minimum purity requirements for water reuse have been specified. Primary and secondary stages were studied considering three types of technologies for the former, and four types of technologies for the latter. In primary treatment, the technologies are based on the particle size: sedimentation, filtration, and microfiltration; whereas, in secondary treatment, the technologies are based on the ability to remove chemicals, such as adsorption, disinfection, and rotating biological contactors, and a tertiary treatment method of advanced oxidation process is used if the desired purity limits are not achieved in prior treatment stages. The contaminant removal efficiencies and costs associated with the operation of these technologies were obtained through information collected from design literature. The primary and secondary treatment stages were studied by means of the P-graph approach, which has been extensively deployed to evaluate problems of combinatorial nature, such as supply chains, evacuation routes design, and process network synthesis (PNS) [4]–[6]. This advantage is that not only the best structure in terms of cost is obtained, but also a ranked list of all feasible near-optimal solutions is created. This information can be relevant when considering additional performance criteria, such as controllability, operability, flexibility, sustainability, etc.

For the primary stage, an initial structure was obtained by joining the input and output materials with the technologies. The water-rich streams leaving the primary stage can be considered as potential input for the next stage. The secondary structure is constructed for each of the water-rich streams leaving the primary stage by joining the plausible materials and technologies of the secondary stage. The resultant overall network consisting of the initial and secondary structure is used as the input for the P-graph algorithm, which determines the most cost-effective treatment process for a certain set of parameters. Subsequently, a parametric study is carried out for a range of feed water compositions, with the objective of generating cost-effective process structures that are also capable of satisfying the treatment requirements under different inlet water conditions. Therefore, this work aims at providing guidelines for the design of efficient wastewater treatment networks depending on the inlet water composition and treated water requirements.

Keywords: Wastewater treatment, process network synthesis, exhaustive enumeration, P-graph

References:

[1] M. & E. Inc, G. Tchobanoglous, F. L. Burton, and H. D. Stensel, *Wastewater Engineering: Treatment and Reuse*, 4th edition. Boston: McGraw Hill Higher Education, 2002.

[2] K. M. Yenkie, "Integrating the three E's in wastewater treatment: efficient design, economic viability, and environmental sustainability," *Current Opinion in Chemical Engineering*, vol. 26, pp. 131–138, Dec. 2019.

[3] K. M. Yenkie, S. Burnham, J. Dailey, H. Cabezas, and F. Friedler, "Generating Efficient Wastewater Treatment Networks: an integrated approach comprising of contaminant properties, technology suitability, plant design, and process optimization - ScienceDirect," in *29th European Symposium on Computer Aided Process Engineering*, Eindhoven, Netherlands, 2019, vol. 46, pp. 1603–1608.

[4] I. Heckl, F. Friedler, and L. T. Fan, "Solution of separation-network synthesis problems by the P-graph methodology," *Computers & Chemical Engineering*, vol. 34, no. 5, pp. 700–706, May 2010.

[5] H. Cabezas, A. Argoti, F. Friedler, P. Mizsey, and J. Pimentel, "Design and engineering of sustainable process systems and supply chains by the P-graph framework," *Environ. Prog. Sustainable Energy*, vol. 37, no. 2, pp. 624–636, Mar. 2018.

[6] L. Vance, H. Cabezas, I. Heckl, B. Bertok, and F. Friedler, "Synthesis of Sustainable Energy Supply Chain by the P-graph Framework," *Ind. Eng. Chem. Res.*, vol. 52, no. 1, pp. 266–274, Jan. 2013.

589912 Robust Superhydrophobic Graphene Quantum Dots Coating Sponge As an Efficient Sorbent for Environmental Oil Spill Treatment

Abstract:

There is an immense need for efficient cleanup and recovery of crude oil spill in society today. Numerous oil spill catastrophes have resulted in the loss of billions of dollars due to the waste of products and treatment efforts, which now has drawn considerable attentions. Newly sorbent materials of excellent oil wettability to remove oil pollutants are highly required. In this study, a robust superhydrophobic graphene quantum dots layer was coated on the treated porous melamine sponge through freeze drying process. The eco-friendly surface functionalized graphene quantum dots sponge with a contact angle of 155.0 degree represented high oil absorption efficiency under various environmental conditions like temperatures, mechanic stress and water pH. The sponge also showed its good recyclability after 50 usage cycles, which develops great potential to work as an efficient oil spill sorbent in severe marine environments.

AUTHOR INDEX

AIChE
120 Wall Street, FL 23
New York, NY 10005-4020

ISBN 978-1-7138-2326-1
9 781713 823261

CURRAN ASSOCIATES INC.
proceedings.com